3000 8
St. Louis Community

D0847911

Cosmic Time Travel

A Scientific Odyssey

OTHER RECOMMENDED BOOKS BY BARRY PARKER

COLLIDING GALAXIES
The Universe in Turmoil

INVISIBLE MATTER AND THE FATE OF THE UNIVERSE

CREATION
The Story of the Origin and Evolution of the Universe

SEARCH FOR A SUPERTHEORY
From Atoms to Superstrings

EINSTEIN'S DREAM
The Search for a Unified Theory of the Universe

Cosmic Time Travel

A Scientific Odyssey

Barry Parker, Ph.D.

Drawings by
Lori Scoffield

Plenum Press • New York and London

Library of Congress Cataloging-in-Publication Data

Parker, Barry R.
Cosmic time travel : a scientific odyssey / Barry Parker ; drawings by Lori Scofield.
p. cm.
Includes bibliographical references and index.
ISBN 0-306-43966-2
1. Time. 2. Time travel. I. Title.
QB209.P35 1991
529--dc20 91-18098
CIP

ISBN 0-306-43966-2

Plenum Press is a division of Plenum Publishing Corporation
233 Spring Street, New York, N.Y. 10013

Printed in the United States of America

Preface

Einstein wrote in one of his last letters, "People like us, who believe in physics, know that the distinction between past, present and future is only a stubborn, persistent illusion." It might seem strange that the man who gave us more insight into the nature of time than any other living being would write that time is an illusion. Is it an illusion? Most people would say, no. They would, however, be hard-pressed to define it. Time is, without a doubt, one of the most elusive properties of the universe, so simple yet so incomprehensible.

Eighty-five years ago Einstein came to the conclusion that time had no independent existence apart from the ordering of the events. He used this in his formulation of the special theory of relativity. Newton had thought of time as inflexible and absolute; Einstein proved that he was wrong. Three years later, Herman Minkowski showed that time was intricately connected with space, part of a space-time continuum. The big breakthrough in our understanding of space and time, however, came in 1915 when Einstein published his general theory of relativity.

Time still mystifies and intrigues scientists. One of the greatest mysteries associated with it is time travel, travel into the past and future. Thousands of science fiction stories have been based on the idea. Many articles and books have discussed it. Still, many misconceptions persist. In this book, I hope to clear up some of the confusion. Time travel will be examined from a scientific point of view. What do scientists think of it? Do they take it seriously?

How well do they understand it? And perhaps the most important question of all: Is time travel possible? These are a few of the questions we will explore.

As in my previous books, I will discuss both the ideas and the scientists who formulated them. Considerable controversy still surrounds time travel and many of the ideas are speculative. Some of them will, no doubt, be proved wrong in the future. But this is expected, it's the way science progresses.

Technical terms have been kept to a minimum, but there may be words that you are unfamiliar with. Because of this, I have included a glossary.

Very large and very small numbers are also needed occasionally. Rather than write them out explicitly, I have used scientific notation. In this notation a number such as 100,000 is written as 10^5 (i.e., the index gives the number of zeros after the one). Similarly, very small numbers such as 1/100,000 are written as 10^{-5}.

I am grateful to the scientists who consented to interviews. Some of them also sent me reprints and photographs. They are: C.T. Bolton, Anne Cowley, Jeffrey McClintock, Michael Morris, John Friedman, Ian Redmount, John Wheeler, Richard Price, Don Page, Sidney Coleman, Andrew Strominger, Steve Giddings, Thomas Banks, Alan Guth, William Unruh, and William Metzenthen. I would particularly like to thank William Metzenthen for his sequence of computer-generated diagrams depicting entry into a black hole.

The sketches, paintings, and line drawings were done by Lori Scoffield. I would like to thank her for an excellent job. I would also like to thank Linda Regan, her assistant Naomi Brier, and the staff of Plenum for their assistance in bringing this book to its final form. Finally, I would like to thank my wife for her support while the book was being written.

Contents

CHAPTER 1

Introduction

Ever since H. G. Wells penned his novel *The Time Machine*, man has dreamed of breaking the bonds of time and traveling into the misty future, and back to the past. What would it feel like to be suddenly catapulted hundreds, even thousands of years into the future? What would it be like to visit your ancestors? Science fiction writers have been speculating on such matters for years. But will it ever become reality? Scientists have recently been looking into this and what they have found has astounded them. Time travel may, indeed, be possible.

We would not only like to travel to the future and the past, but we would also like to travel long distances in the universe—to distant stars. We know, however, that there is no possibility of doing this using conventional methods of travel. Even at speeds of millions of miles per hour, it would still take thousands of years to reach the nearest stars. We somehow have to overcome the vastness of space—perhaps tunnel through it in some mysterious way.

But to do this we have to learn much more about space and time. Do we really know what space is? The answer has to be no. Fortunately, we do have control over space: we can move to any point in it at any time (within our immediate neighborhood). Time, on the other hand, appears to be uncontrollable; it ticks away relentlessly no matter what we do. As much as we would like occasionally to slow it down, we can't. Nor can we speed it up. It passes at the same monotonous rate regardless of anything we do.

The stars in this photograph are so far away it would take millions, even billions of years to get to them using conventional methods of travel. (Courtesy Lick Observatory.)

What is time? Do we really understand it? Newton was convinced that it was absolute, the same for everyone throughout the universe. But, in 1905, Einstein proved that he was wrong. Time could pass at different rates for different observers. If you were traveling at a speed close to that of light relative to someone, his time would appear to run slow compared to yours (see Chapter 2).

This was a significant breakthrough. For the first time it appeared as if our dream of flight to the stars might be possible. Leaving the Earth and traveling at speeds close to that of light, we would be able to travel light-years (a light-year is the distance light would travel in a year's time) within weeks, or even days. But, when we returned to Earth, we would find that thousands of years would have elapsed while we were gone.

Einstein's special theory of relativity opened our eyes to some of the mysteries of time. But it was a restricted theory, applying only to uniform, straight-line motion. And Einstein was not satisfied with this. He wanted his theory to apply to all types of motion—in particular, to accelerated motion. For ten years he struggled to generalize the theory, and, finally, in 1915, he succeeded. But, strangely, his new theory was not just a generalization of special relativity to accelerated motion, it was a theory of gravity. Acceleration and gravity, it turned out, were intricately linked.

This new theory gave an important new insight into the nature of space and time: gravity had a pronounced effect on them. But, because gravity is created by matter, it was really matter that controlled them. Matter curves space—the greater the density of matter, the greater the curvature. Furthermore, the greater the curvature, the slower time runs.

The strangest outgrowth of general relativity, however, was a mysterious "forbidden sphere" associated with a dense region of matter. The interior of this sphere appeared to be cut off from our universe. General relativity could tell us nothing about it. The radius of this strange sphere is now called the gravitational radius.

Scientists were dismayed. What was the significance of this radius? No one knew. Furthermore, it was soon discovered that space was twisted into a bizarre funnel leading down to this forbidden sphere. It was, in a sense, a "tunnel in space." But no one understood it. No one knew what caused it. Then Einstein and a colleague, Nathan Rosen, discovered that the tunnel had a "mirror image" tunnel attached to its other end. It was like a wormhole in space with two entrances. Midway between the entrances was the forbidden sphere.

What would happen if you tried to pass through this wormhole? Where would you end up? Einstein's only answer was that you would end up in another universe, but the idea was repugnant to him. Looking into the details he was relieved when he found it would take a speed greater than that of light to get through it. It was therefore impassable.

But did such things actually exist in nature? Were there forbidden spheres and tunnels in space? If so, how would they arise? Since space was curved by matter, a high density of matter would obviously be needed—densities far beyond those known on Earth. Oddly enough, many years earlier the British clergyman and geologist John Michell had shown that bizarre, dense, stellarlike objects might exist—objects that could not be seen directly. He showed that if a "star" was sufficiently dense and massive, no light would leave it. It would look like a black sphere against a background of stars.

But how could anything become this dense? As astronomers began to understand stars better, they discovered that they went through a life cycle, just as humans do. They are born, live for millions or billions of years, then die. And when they die, they collapse in on themselves. If a star is sufficiently massive it can actually collapse inside its gravitational radius.

Robert Oppenheimer of the California Institute of Technology (Caltech) was the first to use Einstein's theory to study this collapse. And what he found mystified him. The star would collapse forever, but, as it passed its gravitational radius, it would leave a black sphere—a black hole in space. Strangely, though, all the matter of the star would continue collapsing into a point—a "singularity"—at the center of this black hole.

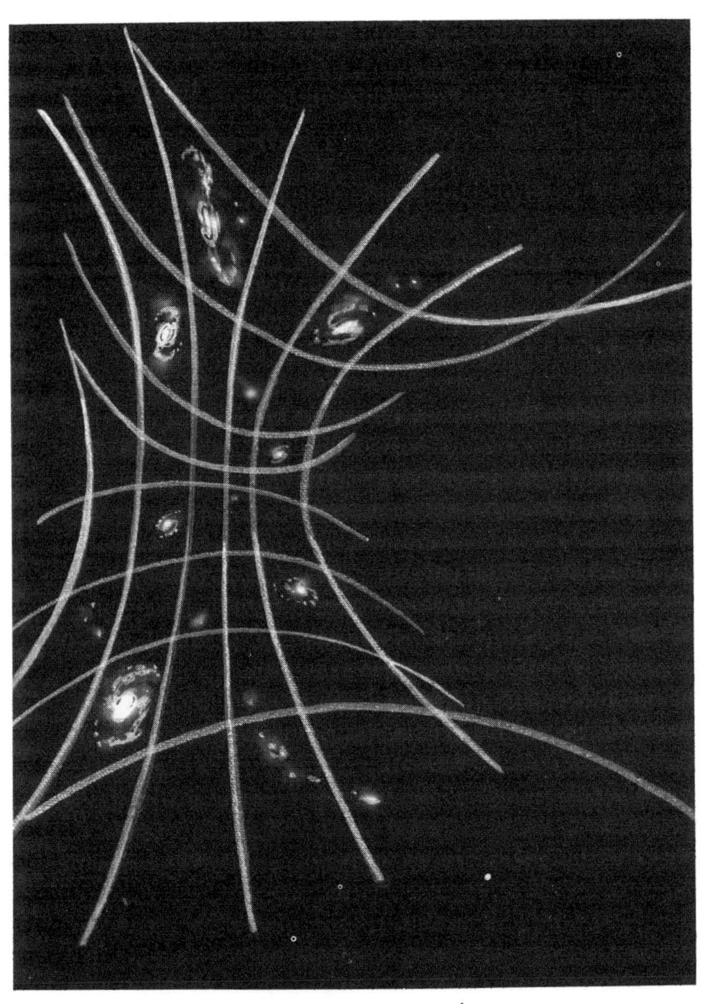

A schematic showing curvature of space.

The surface of a black hole is known as an event horizon. It is a one-way surface. Once inside it, you cannot escape; to get out you would need a speed greater than that of light, which relativity tells us is impossible.

This black hole was named after Schwarzschild, the scientist who first predicted it using Einstein's theory. It has a wormhole associated with it, and if we could pass through this wormhole, we would have a "subway" to distant points of the universe. Furthermore, it could also take us to the future and past. But as Einstein showed, it takes a speed greater than that of light to get through it, so it had little appeal for space travel.

There are, however, other types of black holes. When a spinning star collapses, and, indeed, most stars do spin, it creates a spinning black hole. The solution for this type of black hole was found by Roy Kerr of New Zealand, and it is now named after him. Kerr's solution showed that a spinning black hole is much more complex than a Schwarzschild black hole. It has two event horizons and a strange region just outside them called the ergosphere.

Soon after the solution for this type of black hole was found, scientists checked to see if it was possible to pass through the wormhole that was associated with it. And, indeed, it was. A speed greater than that of light was not needed. Time travel, it appeared, might be possible after all.

We now know that there are four types of stellar-collapse black holes. Besides the Schwarzschild (nonspinning) and Kerr (spinning) types, there are charged black holes (Reissner–Nordström) and charged-spinning black holes (Kerr–Newman). All of these black holes are more complex than the Schwarzschild variety.

Furthermore, black holes do not arise only in the collapse of giant stars. Stephen Hawking of Cambridge University has shown that they might have been created in the big bang explosion that gave us the universe. If this explosion was inhomogeneous, pockets of matter would have been squeezed into black holes. To distinguish them from the stellar-collapse variety, we call them primordial black holes. They are quite different in that they can range considerably in size and mass, all the way from tiny atomic-sized black holes up to gigantic ones that may now reside in the cores of galaxies. All stellar-collapse black holes, on the other hand, are a few miles in diameter.

Scientists soon determined that it was possible to pass through the wormholes of all black holes with the exception of the Schwarzschild type. This generated considerable excitement. Was it possible that they could be used as time machines and allow us to travel to the stars? As the details were worked out, difficulty after difficulty was encountered. First it was found that they were unstable. If a spaceship attempted to travel through one of these wormholes, it would pinch off, crushing the ship and crew. In addition, severe stretching forces would be encountered as you entered the wormhole. They would pull you apart; by the time you got to the mouth of an average stellar-collapse black hole, in fact, you would be stretched into a piece of string. And finally, if you did somehow manage to survive after getting into the wormhole, you would encounter exceedingly high radiation levels. They would be so high, in fact, that you would be literally fried.

The problems were so severe that scientists soon abandoned the idea. As far as they were concerned time travel via black hole wormholes was impossible. This was a great disappointment to science fiction writers, science fiction fans, and even many scientists. For years it seemed as if time travel was beyond the reach of mankind. No one knew a way around the difficulties. Then, in the late 1980s, an important breakthrough was made. Scientists discovered that wormholes that were not associated with black holes (assuming they existed) could be made traversable. The instability, or pinching off, could be prevented by applying a special "balm" to the sides of the wormhole. But this was no ordinary balm; nothing like it was known on Earth. It was so strange, and different, that scientists referred to it as "exotic matter."

When this exotic matter is applied to the wormhole, it cannot pinch off. Furthermore, it was soon found that the wormhole could be designed so that the stretching forces were minimal, and radiation was negligible. And, since there was no event horizon, they could also be made two-way. The key was the exotic matter, but this, unfortunately, was also the stumbling block: we are still not sure it can be produced. If someday we are able to produce it, though, time travel may be possible.

The details of making the wormhole into a time tunnel have now been worked out. All you need to do is move one end of the wormhole at high speed (close to the speed of light) or place it in a strong gravitational field. Either will do the trick. Once this is done, you could go in one end and come out the other at an earlier time. Or you could go in the other end, go through it in the opposite direction, and come out in the future.

It sounds like science fiction, but it is scientifically sound. There is, however, a problem: the principle of causality. This principle tells us that cause has to come before effect. If you throw a rock at a window, for example, the window must break after the

A wormhole in space (artistic conception).

rock is thrown, not before. With time tunnels, this might not be the case. Several scientific groups are now looking into the difficulties associated with causality.

We still have the dilemma of where we would get a wormhole. We want it to be independent of black holes, yet the only place where macroscopic wormholes occur in nature is in connection with black holes. Fortunately, there might be a way around this difficulty. On an extremely small scale, referred to as the Planck scale (10^{-33} cm), scientists believe there are a myriad of tiny wormholes. A considerable amount of work is, in fact, now going on in an attempt to understand these tiny wormholes. Stephen Hawking of Cambridge, Don Page of Penn State, Sidney Coleman of Harvard, and many others are presently looking into various aspects of these wormholes. They believe that a better understanding of them may one day help us attain one of the most sought after theories in physics: a quantum theory of gravity.

If such wormholes do exist, and literally all scientists working in the area believe that they do, they might somehow be expanded up to macroscopic size. We have no idea at the present time how this would be done. But if it were possible, we would have a wormhole that does not have a black hole within it, one that we could stabilize and make into a time machine. It is a fascinating prospect.

If this is possible, we can't help but wonder if other civilizations somewhere out in the universe are using them now. We don't have any direct evidence that other civilizations exist, but statistically it is reasonable to assume that they do. After all, there are 200 billion stars in our galaxy alone, and there are hundreds of billions of galaxies just like ours. The probability that there are civilizations somewhere is overwhelmingly high, and, if there are, it is quite possible that some of them are much more technologically advanced than we are. If so, they may be using time travel.

We have learned much about time and time travel in the past few years, but there is still much more to be learned. In the early chapters of this book we will look at the history of the theory that started it all—the general theory of relativity. We will see that it

Passing through a wormhole to another universe.

The probability that there is life somewhere among the stars is high according to most astronomers. This photograph shows several galaxies. (Courtesy National Optical Astronomy Observatories.)

predicts both black holes and wormholes. In the latter part of the book we will investigate the latest advances in the area and see that time travel may, indeed, be possible. We will then consider the possibility that supercivilizations exist, civilizations that now may be using time tunnels. And, finally, we will take a brief look at another possibility for traveling to the stars—using superluminal speeds, speeds greater than that of light.

CHAPTER 2

Einstein and the Elasticity of Time

Over the last few years man has conquered much of the solar system; he has visited the moon and sent probes to all of the planets with the exception of Pluto. But, as he pushes farther into space, the challenges and obstacles mount. It took Apollo only three days to reach the moon, but Pioneer 10 took nearly two years to reach Jupiter, and it was 14 years before it finally exited the solar system. At this speed it would take 80,000 years for us to get to the nearest star, Alpha Centauri, which is 4.3 light-years away, and Alpha Centauri is of little scientific interest. More promising are stars such as Tau Ceti, which is at a distance of 11 light-years and Zeta Reticuli, which is 30 light-years away. But, unless we can increase our speed of travel considerably, we are going to have problems visiting them, or any other stars for that matter. Furthermore, when we consider the speeds that are needed for efficient space travel, we encounter another problem: the force on our body due to the acceleration required to reach these higher speeds would quickly overcome us. The human body can only stand 2 or 3 g's (g is the acceleration of gravity on Earth) over a long period of time.

There is, however, a way around part of this problem, a way that was pointed out by Einstein in 1905. In his special theory of relativity, Einstein showed that time does not pass in the same way for everybody. If someone moves relative to you at a high speed (the effect is greatest close to the speed of light), his clock will run slow compared to yours. Space travelers of the future may be able to take advantage of this effect.

Einstein's ideas are now accepted by literally all scientists. But do they really apply to space travel? For years people found it difficult to believe that all functioning clocks do not run at the same rate. It seemed preposterous that an astronaut could fly off into space at a speed close to that of light, then come back to Earth many years in the future and have aged only a few months. But this is exactly what Einstein's theory predicted.

Prior to Einstein's discoveries, scientists were convinced that time was absolute. In other words, it was the same for all observers throughout the universe. The English physicist Sir Isaac Newton was the first to suggest this. In his famous book *Principia,* which was published in 1687, he wrote, "Absolute, true and mathematical time, of itself and from its own nature, flows equably, without relation to anything external, and by another name is called duration." The crucial words here are, "without relation to anything external." Newton was obviously convinced that time was independent of any clock or any observer. He felt the same way about space; referring to it he wrote, "Absolute space in its own nature, without relating to anything external, remains always similar and immovable." Space, like time, according to Newton, was absolute. And Newton's influence was so strong in the early years following the publication of *Principia* that no one challenged him. But, in time, a number of scientists began to feel uncomfortable with the idea. In fact, in his later writings it became obvious that even Newton himself was not entirely satisfied with it.

It is, in fact, easy to see a flaw in the idea of absolute space. We know that the size of everything in the universe is determined by comparing it to a measuring rod. But what if everything, including the measuring rod, suddenly shrunk in size? We would never know a shrinkage had occurred, since everything would appear the same to us.

A similar argument can be given for time. Furthermore, if time were absolute, we would be able to set all clocks throughout the universe to the same time. The instant "12:00 noon" would be the same for all observers. But if we wanted to check with someone on, say, Alpha Centauri, to see if their clocks were synchronized with ours, it would take 4.3 years to get a message to them and another 4.3 years to get a reply. It seems, on the basis of arguments like this, that neither space nor time can be absolute.

A serious problem, however, arose when James Clerk Maxwell of Cambridge University showed that light was an electromagnetic wave. If light was indeed a wave, a medium of some sort was needed to propagate it. To understand why this is so, consider the following scenario: assume you throw a rock into a pond. What happens? Waves, of course, move across the surface of the pond. But what happens if you take away the water? Obviously, you don't get a wave; water is needed for the wave to propagate. In the same way a propagating medium is needed if light is to pass through space. Since we knew that light traveled to us from the sun and stars, scientists assumed that there had to be a medium filling all of space. They called it "aether." This aether, however, had to have some amazing properties; we couldn't see it directly so it had to be transparent, and to transmit waves over long distances it had to be incredibly rigid.

But once scientists had invented the aether they realized they had another problem. If the universe was permeated with an aether, it would act as a "frame of reference." This meant that we would be able to measure the speed of everything that moved through it, in the same way that we are able to measure the speed of a boat relative to the water it moves through.

Our sun was presumably moving through the aether, so its speed could be determined relative to it. How? It would be difficult to measure its speed directly, but, since the Earth was going around the sun, we could get it indirectly by measuring the Earth's motion through the aether.

In 1881, an American scientist working in Italy named Albert Michelson set up an experiment to see if he could detect the Earth's motion through the aether. He projected a beam of light in the direction of the Earth's motion and one opposite its motion. In the first case, the Earth would be catching up with the light beam so it would appear to be moving away from us at the recognized speed of light (186,284 miles per second) minus the speed of the

Albert Michelson.

Earth in orbit (approximately 20 miles per second). It would be similar to having someone shoot a bullet past you while you are in a jet. If you measured the speed of the bullet from the jet, it would obviously appear to be going slower than it would be if you measured its speed from the surface of the Earth. Michelson expected the same effect in the case of the light beam, but was surprised to find that the Earth's speed had no effect on the speed of the light beam. Regardless of whether the light beam was projected in the direction of the Earth's motion or in the opposite

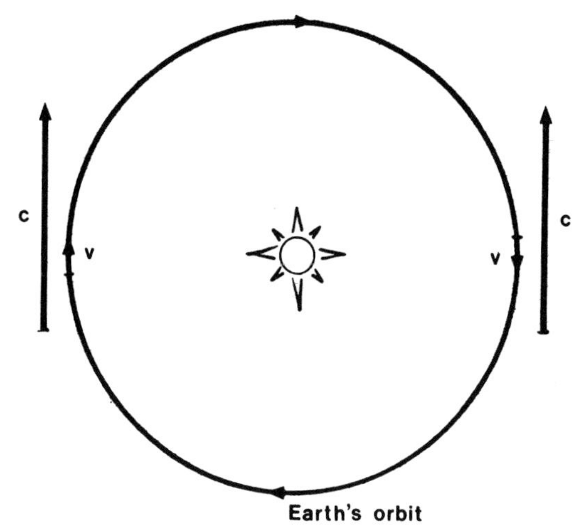

The Michelson–Morley experiment. If a beam of light (speed c*) is projected through the aether in the direction of the Earth's velocity* (v)*, the Earth should appear to catch up with it. Its speed would therefore appear to be* c − v*. This is shown on the left-hand side. On the right-hand side we have the Earth and the light beam moving in opposite directions. The speed of light in this case should appear to be* c + v*. Michelson and Morley showed that this is not the case.*

direction, the speed of light was always measured to have the same value (approximately 186,284 miles per second).

He repeated the experiment in 1887 with a colleague, Edward Morley, and obtained the same result. Scientists were shocked. Most didn't know what to make of the strange result. Edward Fitzgerald, in Ireland, suggested that the aether was exerting a strong pressure on the measuring rods and as a result they were shrinking in the direction of motion. He derived a formula for the amount of shrinkage. H. A. Lorentz, in Germany, derived the same formula independently, but considered the contraction to be a result of an electromagnetic force produced by the motion through the aether. He assumed that the atoms were moving closer together as a result of this force.

For several years scientists were at an impasse. Nothing seemed to make sense. Then, in 1905, a paper was published that offered a simple and logical answer; it was so simple and logical, in fact, that scientists were reluctant to accept it at first. Albert Einstein, working in a patent office in Bern, Switzerland, showed that if time was not absolute—not the same for all observers, but dependent on their relative motion—the pieces would fall into place; everything would make sense. The idea was revolutionary, but the important thing is that it worked. It explained the experimental results beautifully; furthermore, it told us that flight to the stars might one day be possible.

How was Einstein able to make such an astounding prediction? To understand his work, it is best to start with a look at his life.

EINSTEIN: THE MAN

Einstein was born in Ulm, Germany on March 14, 1879. His father, Herman, was a small-factory owner, and his mother, Pauline, a housewife. A year after he was born, the family moved to Munich where a daughter Maja was born.

Though Jewish, Einstein attended the Catholic school in Munich from 1884 until 1888 when he entered the gymnasium—the equivalent of our high school. He was a good student at the Catholic school, but he doesn't seem to have impressed his teachers. His father once asked one of them what profession his son should go into and was told, "It doesn't matter; he will never make a success of anything, anyway."

Einstein's interest in mathematics began early, and it was at least partially due to an uncle who lived with the family. Herman's brother Jacob gave Einstein books to study and encouraged him to solve the problems in them. By the age of ten, Einstein had proved many difficult theorems, including Pythagoras's theorem.

About 1890, however, Herman's and Jacob's business began to fail and Jacob spent less and less time with Einstein. A new

Albert Einstein with his sister Maja. He is about five years old. (Courtesy Lotte Jacobi.)

"teacher," however, soon came into his life. The Einstein family was not a religious one, but they kept an old religious tradition: once a week they shared a meal with someone of meager means. One of the recipients of their generosity was a 21-year-old medical student named Max Talmey. Talmey was interested in science and math, and soon took a liking to young Einstein and began having long talks with him. He brought him several popular science books which Einstein eagerly devoured. Talmey soon found, though, that Einstein's mind worked so fast that it was difficult to keep up with him. He finally gave up trying to talk to him about science and math, and began switching the discussions over to philosphy.

When Einstein was about 12, he came across a geometry book. Fascinated by it, he worked every problem in it, showing his solutions proudly to Talmey. He later referred to it as his "holy geometry" book. It had a strong influence on him, and by the time he had finished it, he was eager to tackle something more difficult. Calculus was the logical next step, and by 14 or 15 he had mastered it. By this time he was far ahead of his classmates in mathematics.

But, unfortunately, math and science were not the only subjects taught at the gymnasium. Latin, Greek, and history were also on the curriculum, and Einstein found them to be of little interest. He therefore spent little time studying them, much to the dismay of his teachers. More than anything, though, he hated the militarism of the school—the forced drills and the strict discipline.

The family business continued to go downhill, and, in the spring of 1894, Herman decided to move to Milan, Italy and start anew. Einstein was excited, but when his father told him that he would be left behind in Munich to finish his schooling he was distraught. He was sent to board with a woman he hardly knew. It was a trying period for him; he was lonely, depressed, and soon began to neglect his schoolwork. His resentment toward the school and the curriculum had always been there, but now it began to show.

Finally, he decided he'd had enough. He was going to join his family in Milan at any cost; he devised a plan. First, he went to

Max's brother, a medical doctor, and persuaded him to give him a medical certificate saying that he was close to a nervous breakdown and needed a rest in Italy. He then approached his mathematics teacher and asked for a letter stating that his mathematical education was well beyond that required for the gymnasium. To his surprise the teacher gave him one.

But there was also a surprise that he was not expecting. As he got ready to leave, one of his teachers took him aside and asked him to leave the school. The teacher felt he had become a bad influence on the rest of the students. Einstein was dismayed, but at the same time he was relieved—he was finally leaving.

As he took the train to Milan, he was apprehensive about what his parents would say. And, as expected, there were mixed emotions when he arrived. His parents were glad to see him, but they were disappointed that he had left school. Einstein, however, made the promise that he would take the entrance exams at the Polytech in Zurich, and this seemed to satisfy his father. He also shocked his family by announcing that he was going to reject his German citizenship. This was probably done to avoid having to serve three years in the German army.

For the next several months, Einstein led a carefree existence. A burden had been lifted from his shoulders and his spirits soared. He hiked in the nearby mountains and spent countless hours poring over popular science books. He became particularly close to his sister Maja, and would tell her of his new discoveries and ambitions. In the evening he would look up at the stars and wonder about them. The rays of light that came from them had to have traveled for hundreds, perhaps thousands of years through the aether—that strange, invisible substance that presumably permeated the universe. What effect did this aether have on the light? What was it really like?

He began to wonder what it would be like to ride a beam of light from a star—to travel along with it. But common sense told him that if he traveled with it—at the speed of light—it would cease to be a wave. Furthermore, he wondered what would happen to the beam if one moved at speeds greater than the speed of light. According to Newton's theory such speeds were possible. There were many unsettling problems about light and the aether, and he was determined to understand them.

Einstein's thoughts in this area were outlined in an essay he sent to an uncle at this time. Uncle Caesar, his mother's brother, had bought him a steam engine when he was younger, and had encouraged him to study science. The essay was titled, "Concerning the Investigative State of Aether in Magnetic Fields." It is evident from the paper that he had thought about the problems of fields and the aether in considerable detail, even though he was only 16 years old. In the paper he discussed the nature of the electromagnetic field and pointed out that very little was known about the relationship between the field and the aether. He ended the paper with a program for investigating some of the problems.

It was during this period that his uncle Jacob said of him, "You know it's fabulous with my nephew. After I and my assistants had been racking our brains [on a problem] for days that young sprig got the whole thing in scarcely 15 minutes. You will hear from him yet."

However, once again his father's business began to fail and Einstein was told he would have to go to Zurich as soon as possible and take the entrance exams at the Polytech. His father wanted him to become an engineer. So, in the fall of 1895, he traveled with his mother and took the exam. He was examined in math, physics, biology, and languages and, although he did extremely well in math and physics, he did poorly in biology and languages. His overall score, in fact, was not high enough to allow him to enter. It was a shock to him, but he later said he had no one to blame but himself. He had not prepared for anything but math and physics. Furthermore, the promise he had made to his father about becoming an engineer weighed heavily upon him. He wasn't really interested in pursuing a career in engineering.

The director at the Polytech noticed that he did exceptionally well in both math and physics, and arranged for him to go to a high school at nearby Aarau. Einstein stayed with one of the teachers, and soon felt like one of the family. This time he enjoyed the school; it was completely different from the schools at Munich. Gone were the drills and the militarism. Free thinking was encouraged and Einstein loved it. Furthermore, he had considerable spare time to study scientific books. Many of the books were mathematical and went far beyond the curriculum at Aarau. It was a critical period in his scientific development, perhaps even more important than his later years at the Zurich Polytech.

The basic problems of physics were never far from his mind. What was the aether really like? How did it affect electric and magnetic fields? He felt compelled to find out.

In the fall of 1896, he graduated from Aarau and was ready for the Polytech at Zurich. Enrolling as a student of physics and mathematics, he planned on becoming a teacher. The four years he spent at Zurich provided him with a basic foundation in physics and mathematics. But, as strange as it might seem, he did not take advantage of the extensive opportunities that were available to him. Although he took a large number of math classes, he rarely attended lectures. Most of the time he could be found studying on his own or working in the laboratory. It might seem odd that he would spend a lot of time in the laboratory; after all, his real interest was theoretical physics. It is true that his first love was theory; nevertheless, experimental physics intrigued him. He performed many experiments related to electric and magnetic fields, and even designed one to determine the effect of the aether on the fields. His teacher, Wilhelm Weber, however, would not allow him to carry out the experiment on the aether. Weber, in fact, seemed to discourage his work in the laboratory. Einstein, in turn, eventually developed a distaste for Weber's lectures and began skipping many of them. His relationship with Weber soon became quite strained. The final straw came when Weber failed to discuss Maxwell's electromagnetic theory in his course on electricity and magnetism.

"Einstein, you are clever," Weber once said to him. "But you have one fault; no one can tell you anything." Furthermore, he was also not in the good graces of his math teacher, Herman Minkowski. Minkowski once called him a "lazy dog."

Einstein as a student at Zurich Polytechnic Institute. He is in his early twenties. (Courtesy Lotte Jacobi.)

Although he attended lectures only occasionally, he devoted much of his time to self-study. He studied the works of the pioneers of electricity and magnetism, men such as Helmholtz, Hertz, and Maxwell. Nevertheless, his neglect of regular classes finally caught up with him, and he had to cram extremely hard for his final exams. Luckily, a classmate, Marcel Grossman, had taken excellent notes and he loaned them to Einstein.

So, in the spring of 1900, with the help of Grossman's notes, Einstein graduated from Zurich Polytech. Years later he remembered the last few weeks of cramming as a distasteful experience. He said it exhausted him so much he that couldn't think seriously about physics for many months.

Upon graduation he hoped to get a job at the Polytech as an assistant to one of the professors. He went around to each of them asking to be considered, but found that no one wanted him. He was even turned down by his physics teacher. It was a harsh blow and left him greatly disillusioned. He began applying for other jobs, but was unsuccessful. Late in the year he published his first scientific paper in *Annalen der Physik*; it was on capillary action. He had become interested in capillary action after reading the works of the Dutch physicist, Friedrich Ostwald. Hoping that Ostwald might be impressed with the paper, he sent him a reprint of it, asking if there were any positions available in his laboratory. He got no reply. Unknown to Einstein, his father also wrote a letter to Ostwald, but nothing came of it either.

In February of 1901, he obtained Swiss citizenship and presented himself to the military for 6 months compulsory service. Although he had previously rejected his German citizenship to avoid serving in the German army, he now found himself looking forward to serving in the Swiss army. To his surprise, though, he was rejected because he had flat feet and varicose veins. His self-esteem sunk even lower. Finally, in the spring he found a temporary teaching job. He was jubilant, but within a few months he was on the streets again looking for a job. He wrote to his old friend Marcel Grossman from the Polytech, telling him of his predicament. Grossman talked to his father, who in turn talked to the director of the patent office in Bern. Grossman wrote back telling Einstein that he would be seriously considered when the next position became available at the patent office.

In the spring of 1902, there was an opening and Einstein was hired as a technical expert, third class. He began work on June 23. His salary was small, but he was happy. He enjoyed the work and it left considerable time for him to continue his study of theoretical physics. Furthermore, he found that looking over the numerous inventions he had to examine stimulated his mind.

His life was finally beginning to turn around. Then tragedy struck. His father's business in Milan failed again and the strain was too much for him. He died on October 10. Einstein was devastated; he later wrote that it was "the deepest shock he had ever experienced."

When he returned to Bern he began to think about marriage. He now had a full-time job and felt that he could support a wife. While at Zurich Polytech he had met a Slav girl by the name of Mileva Maric. They had talked of marriage, but, because she was a Slav, Einstein's family was strongly against it. His mother never did give her approval, but on his deathbed, his father relented. So, in early 1903, Einstein married. The couple moved into a small apartment, but before long it became obvious that his small salary was barely enough to live on, so he had to supplement it. Fortunately, Bern was a college town, and there were always students who needed tutoring. Einstein therefore advertised as a tutor in math and physics and soon had two students: Maurice Solovine and Konrad Habicht. He enjoyed working with the students, but, strangely, he seldom lectured to them. The classes were discussion sessions where the students talked as much as he did. In many ways they were sounding boards for his ideas. But they learned a lot, and both students remained lifelong friends of Einstein. They referred to themselves as the "Olympic Academy."

The discussions were extremely helpful in Einstein's development. They did much to sharpen and finely tune his ideas. By the time the group dissolved he was ready to take on some of the most difficult problems in physics. But, with his two students gone, he needed a new sounding board and he soon found one—a fellow worker at the patent office by the name of Michele Besso. Einstein had met him several years earlier at Zurich Polytech, and had later helped him get a job at the patent office.

THE DISCOVERY OF TIME DILATION

The year 1905 ranks as one of the most important in the annals of physics. The world of physics was ready for a revolution—and it came. Several important discoveries had been made in electricity and magnetism, but many contradictions and enigmas were still waiting to be resolved.

Einstein had published half a dozen papers by this time, but none was as important as those he would publish in 1905. His first paper came early in the spring. It dealt with the effect of light on atoms in metals—what was known as the "photoelectric effect." Sixteen years later it would win him the Nobel Prize. This paper was barely in the mail when he completed another paper, which he submitted to the University of Zurich as a Ph.D. thesis. It was titled, "On a New Determination of Molecular Dimensions." In July of 1905 it was accepted.

Another paper was written in May; this one was on Brownian motion—an irregular zigzag motion of small particles caused by the vibration of atoms. Then came his first paper on relativity, and with it the discovery of time dilation. For years Einstein had been puzzling about the problem of electric and magnetic fields and the effect of motion on them. Suddenly in mid-May, the breakthrough came. For weeks he had been discussing the problem with Besso at the office. One night while walking home with him, Einstein confided that he was ready to abandon the problem; the difficulties, he said, were insurmountable. Then that night, perhaps because the strain of weeks, even months of hard thinking had suddenly ceased, everything came together. He awoke in the

Einstein in 1905, the year he published his special theory of relativity. He is 26 years old. (Courtesy Lotte Jacobi.)

morning with the answer. Rushing to the office he told Besso of his discovery.

The key, he decided, was time. Newton's concept of absolute time had been accepted for about 200 years. Einstein, however, was convinced that time was not absolute. Different observers in different states of uniform motion would not necessarily see time pass in the same way. Furthermore, not even space was absolute; the only absolute of nature, as far as he was concerned, was the speed of light. The speed of light is always the same regardless of the motion of its source. This is, of course, what Michelson and Morley had found in their experiment of 1887, although it is not certain whether Einstein knew of this experiment when he was formulating his theory.

A moment's reflection shows us that Einstein's assumption was reasonable. Consider a binary (double) star system. Assume that the two stars are orbiting one another in a plane that is edge-on to us. At any time one star will therefore be moving toward us and the other away from us. If the speed of light is not independent of the motion of the source, the light beam from the star that is moving toward us will have a greater speed than the one from the star that is moving away from us. In fact, as the two stars move around in their orbit, we would expect to see beams of light with many different speeds. This would make it difficult to distinguish the two stars. But we know that we can distinguish them. In fact, a few years after Einstein's prediction the astronomer Willem de Sitter showed that the light rays coming to us from two stars of a binary system, with one approaching and one receding, had exactly the same speed.

The constancy of the speed of light was Einstein's first postulate. He referred to a second postulate as the "principle of relativity." It stated: There is no way to tell whether an object is at rest or in uniform motion relative to the fixed aether. He went on to say that because of this, the aether was superfluous and could be discarded.

Einstein titled his paper, "On the Electrodynamics of Moving Bodies." It was published in *Annalen der Physik*. In many ways it was a strange paper. He made no reference to earlier work; in particular, he did not mention the Michelson–Morley experiment. But he did refer to Besso in the final paragraph, thanking him for several valuable suggestions.

Einstein hoped to get an immediate reaction to the paper; he was sure it would be a criticized, but he was ready to defend it. The next five issues of the journal, however, contained no mention of it. He began to worry, but then he got a letter from Max Planck of the University of Berlin asking for some clarifications. He was relieved and pleased: Planck was one of the leading scientists in the world at that time.

Despite the fact that the paper was "different," it is now considered to be one of the most important physics papers ever published. It cleared up the problems related to the Michelson–Morley experiment and it resolved the enigmatic relationship between the electric and magnetic fields by showing they were closely related. In doing so, though, it introduced several strange concepts—some so strange that many scientists refused to believe them for years. One of the strangest was length contraction. Objects traveling at high speeds relative to a fixed observer shrunk in the direction of motion. If, for example, a rocketship were to pass overhead at a speed close to that of light, it would appear considerably shortened in the direction of travel. Strangely, though, if you were aboard the rocket, it would appear completely normal. Lorentz and Fitzgerald had, in fact, derived a formula for this contraction several years earlier, but they had not understood the physical significance of their result.

Einstein found that the mass of an object also increased relative to a fixed observer as it approached speeds close to that of light. Actually, there is a change at any speed, but it is only significant when the difference in speeds is close to that of light. At 70 percent the speed of light, for example, the mass of an object only increases by 40 percent. At 99 percent, on the other hand, it increases by a factor of 7.

As far as travel to the stars is concerned, though, his most important discovery was time dilation. He showed that a clock in

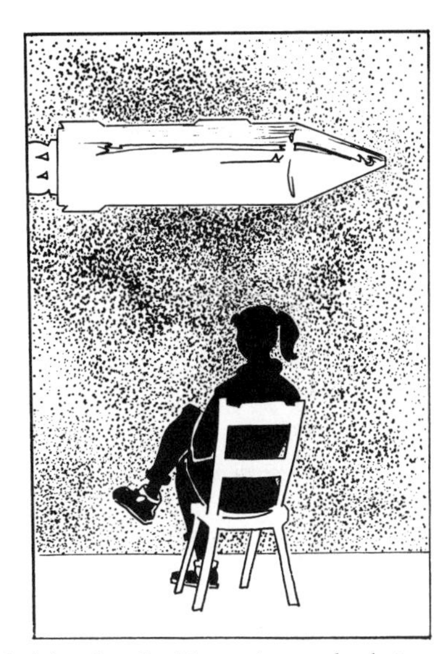

Length contraction as predicted by special relativity. A rocketship passing overhead at a speed close to that of light will appear shortened compared to when it is at rest relative to the observer.

motion would run slow relative to a fixed one (as seen by an observer at the fixed clock), and the faster the moving clock moved the slower it would run. How, we might ask, was Einstein able to make such a discovery? Why did he make it and not someone else? He was asked this question many times throughout his life, and it was as much of a puzzle to him as it was to others. He was, however, convinced that it was at least partially due to the fact that he never lost his childlike awe of nature.

On one occasion he replied to the question by saying, "The reason [I made the discovery], I think, is that a normal adult never stops to think about problems of space and time. These are things he thought of as a child. But my intellectual development was retarded and I began to wonder about them when I was grown up."

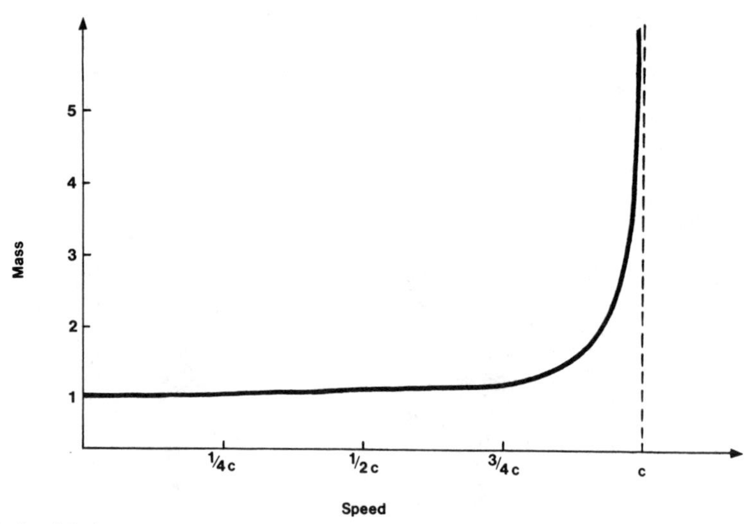

*A plot of the increase in mass with speed. Note that there is little difference at low speeds, but beyond 0.75*c *the increase is significant.*

His reverence and awe of nature was, no doubt, also a factor. One of his most famous quotes is, "The most beautiful experience we can have is the mysterious. Whom does not know it and can no longer wonder, is as good a dead, and his eyes are dimmed."

Another of his traits, one that got him into considerable trouble with his teachers, was that he accepted almost nothing without proving it to himself. He had to reason things out, and understand them, before he would accept them. And he recognized numerous problems surrounding the concept of the aether. His intellectual independence was no doubt a critical factor in his discovery, but along with it he had the ability to concentrate long and hard on a problem. He worked so hard the last few weeks before the completion of his special theory of relativity, in fact, that it left him exhausted and ill. But, of course, not everyone with these traits will necessarily make tremendous breakthroughs.

Einstein had something else—namely the ability to see things as no one else had ever seen them before. That was his genius.

RELATIVISTIC TIME TRAVEL AND THE TWIN PARADOX

Let's turn now to the details of time dilation. To understand how it takes place it is best to begin by assuming you are an observer on Earth watching an astronaut pass overhead in a spaceship. Assume that before the astronaut gets into his spaceship you synchronize your clock with his. Then he gets into his spaceship and flies overhead at a speed close to that of light. If you could see his clock, you would notice that it was running slow compared to yours. Strangely enough, if he could look at your clock, he would also see that it was running slow compared to his. The difference in rates between the two clocks depends on how fast he flies. At speeds close to that of light the difference is significant. But at the speeds we are accustomed to, even space flights such as Apollo, the difference is negligible.

Now suppose that you get into a spaceship and travel alongside the astronaut's spaceship. How do your clocks compare? You will see that they are both running at the same rate. This means that time dilation occurs only when the other observer is moving relative to you.

But if your clock runs at the same rate as the astronaut's when you join him, it seems that there is no advantage to time dilation. How would it be possible for us to use it to travel to the stars? To understand the advantage it is best to think in terms of an experiment involving two 25-year-old twins. Assume that one of the twins stays on Earth and the other gets into a rocketship, travels to a nearby star, and returns. Assume further that the time for the trip according to the twin in the rocketship is one year. How much time passes back on Earth during this time? If the astronaut's speed was 50 percent that of light (written as $0.5c$, where c stands for the speed of light), 1.2 years would pass back on Earth—not a significant difference. On the other hand, if his speed was $0.99c$, seven years would pass back on Earth. And if it was 0.999c, 22 years would pass back on Earth, and the twin that remained on Earth would be 47 years old. His brother, the astronaut, would only be 26.

You have no doubt noticed that I have selected speeds very close to that of light, but have not equaled or exceeded that speed. There is a reason. According to relativity, the speed of light is an upper limit for matter in the universe. You can approach it as close as you want but you cannot equal it. Looking back at some of the relativistic effects that occur, we can see why. Earlier, I mentioned that mass increases relative to a fixed observer as you approach the speed of light. Indeed, at the speed of light, mass becomes infinite. This means it would take an infinite, and therefore impossible, amount of energy to attain the speed of light. Furthermore, we saw that things shrink in the direction of motion relative to a fixed observer at high speed. Again, at the speed of light you get a strange result: objects shrink to nothing! On the basis of this, it should be obvious that we cannot travel at the speed of light relative to another observer.

When the effects of time dilation were first realized, there was considerable controversy. People argued that it couldn't possibly have anything to do with humans. But we know now that time dilation applies to any clock regardless of its construction, and the heart is essentially a clock. Therefore, time dilation must apply to humans. Virtually all scientists are now convinced of this.

Indeed, a number of experiments have verified it. Some of the earliest experiments involved the lifetimes of elementary particles. Particles called mesons are generated in our upper atmosphere when high-speed cosmic rays from space strike the molecules of air in this region. We can measure their lifetime and we have found that it depends on their speed. Comparing their lifetime to that of the same particles produced in the laboratory at low speeds, we find a significant difference—and the difference is in accord with Einstein's theory (called special relativity).

A more direct test was made by Joseph Hafale and Richard Keating of the U.S. Naval Observatory in 1971. They placed an atomic clock aboard a jet and circled the Earth with it. Then they compared the lapsed time with a similar atomic clock back in Washington. Again Einstein's results were verified.

Let's return to the twins we talked about earlier. Some people refer to the fact that the twins age differently, and are not the same age when they get back together, as the twin paradox. But if you think about it for a moment you'll realize that there is actually a much more startling paradox. Let's call our twins Pat and Mike; assume again that both are 25 years old. Suppose Mike makes a trip to a distant star and back traveling at a speed of, say, $0.998c$. It takes him one year. When he gets back from the trip he jumps out of his spaceship, runs up to Pat and says, "Ha! Ha! I'm much younger than you now. I'm only 26 and you're 40."

Pat looks at him in surprise. "No, I'm 26 and you're 40," he says. "According to Einstein, all motion is relative. It was actually the Earth that moved off into space and returned to this position. You just sat here in your spaceship." Mike scratched his head; Pat did indeed have a point. If all motion is relative, then the two situations were similar. Which of the twins was actually younger when they got back together? We now know that one is indeed younger than the other. But it took general relativity, an extension of special relativity which came ten years later, to resolve the difficulty. We'll explore this issue later.

FOUR-DIMENSIONAL SPACE-TIME

It's ironic that a few years after Einstein published his special theory of relativity, an important modification was made to it by one of his old teachers from Zurich Polytech. In fact, it was the math teacher who called him a "lazy dog." Herman Minkowski looked over Einstein's theory and noticed that it could be formulated more elegantly if space and time were not considered to be independent, but rather two parts of what he called "space-time." In 1908, Minkowski delivered a lecture at Cologne that began with the statement: "The views of space and time which I wish to lay

Herman Minkowski.

before you have sprung from the soil of experimental physics, and therein lies their strength. They are radical. Henceforth, space by itself, and time by itself, are doomed to fade away into mere shadows, and only a kind of union of the two will preserve an independent reality."

At the time, interestingly, Minkowski did not really appreciate Einstein's contribution. He said that he was quite surprised that Einstein had been able to formulate such a complex theory.

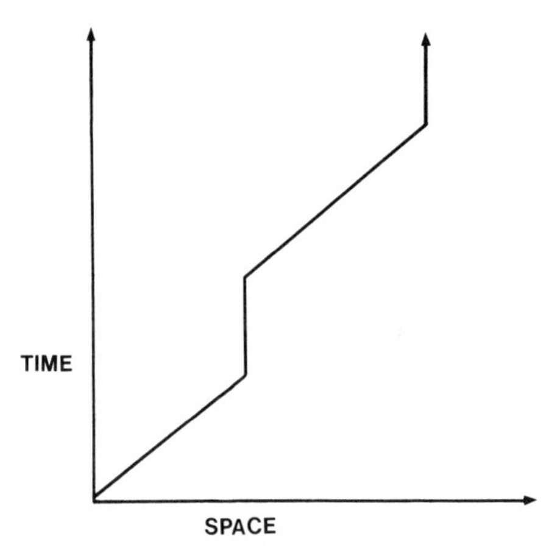

A simple space-time diagram. Plot of a trip between two towns.

On the other side of the fence, Einstein was also unimpressed with Minkowski's contribution; he thought it was of little value. Within a few years, however, he realized it was an important contribution.

To understand Minkowski's contribution it is best to begin with a simple plot of time versus distance. We can represent two events on this graph by two points. The space-time distance between them is just the length of a line drawn between them. If this represented the trip between, say, two towns, the line between the two points would, in practice, be jiggly. In this case it is referred to as the "world line" of the trip.

One of the things we must remember in drawing such a graph is that we can't travel at a speed greater than that of light. We therefore scale the graph so that a line at 45 degrees to the two axes represents the speed of light. Drawing in the two lines at 45 degrees we have two regions that we cannot visit; they are referred to as "elsewhere." A world line will come from the bottom of the graph, pass through the center point (referred to as "now"), and

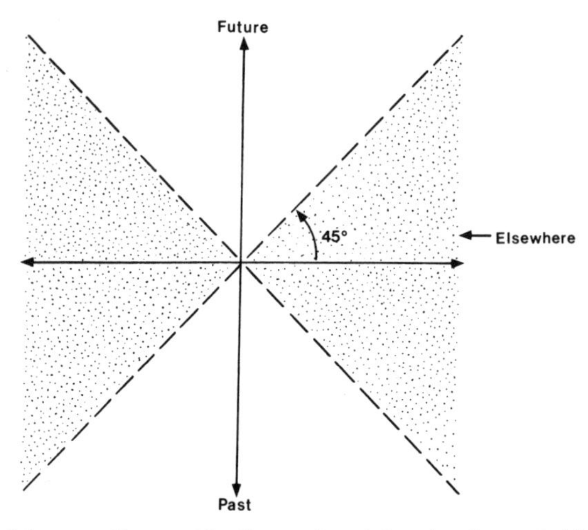

Special relativity space diagram. The diagram is scaled so that the speed of light is at 45 degrees. A space traveler can travel only in the clear regions.

continue upward into the future. It cannot penetrate into elsewhere.

Now suppose we want to extrapolate this to our world of three dimensions plus another dimension, that of time. How do we do this? It turns out we can't do it properly. All we can do is rotate the above diagram around the time axis; the future and past will then be represented by cones. A trip within the upper cone at an angle of less than 45 degrees to the time axis (see diagram) is referred to as "timelike." It is a possible trip. A trip at an angle greater than 45 degrees, on the other hand, is impossible, and is referred to as "spacelike."

Let's consider an interval within this space-time. We'll begin by assuming a supernova (exploding star) is seen by two observers. If they are at different distances from the explosion, they will see it at different times. One may see it at 3:00 PM and the other at 6:00 PM. This difference occurs because of the time it took for the light from the explosion to reach them. On the basis of this, it is

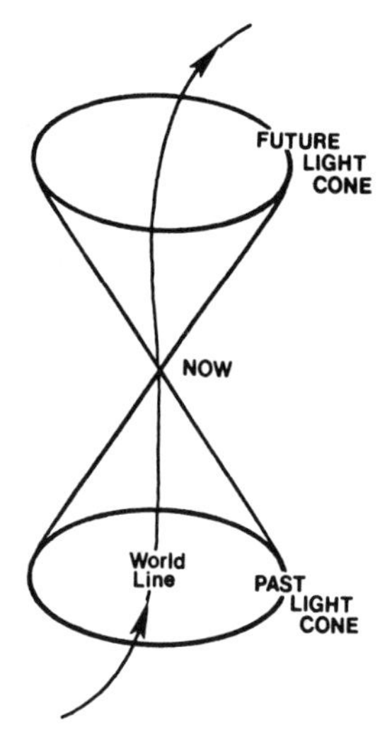

The previous space-time diagram spun on its time axis to give a three-dimensional representation.

easy to see that an interval of space by itself, or time by itself, is not absolute. They depend on the observer—his position and motion. We can, however, bring these two intervals together so that we have an interval that is the same for all observers. To do this, we simply multiply the time by the speed of light to make it into a distance, then add it in the appropriate way to the three dimensions of space. The result is a space-time interval.

One of the major implications of this linking of space and time is that if space somehow became distorted, time would be effected. In other words, if you could somehow bend or twist space, you could change the rate at which a clock runs. It sounds crazy, but it's true. In the next chapter, we'll consider the possibility that space can, indeed, be twisted.

CHAPTER 3

Space That Bends and Twists

If we hope to send rockets to the stars someday, we must have a thorough understanding of space. After all, space is what separates us from the stars. One of the major reasons we want to understand space is because we would like to travel between two points in the universe without actually traversing the space between them. Impossible, you say. Yes, it does sound impossible, but we will see in this chapter that we might eventually be able to do it.

WHAT IS SPACE?

Ask someone to describe space to you and they're likely to say, "What's there to describe? It's just the emptiness around us." And of course this is true. But, as we will see, it's not the entire story—there's actually much more to space than that. One of the first things you learned about space in school is that it is three-dimensional; in other words, it has height, width, and depth. This means that if you take matchsticks and try to set them up perpendicular to one another, you can only make three of them mutually perpendicular. No matter how hard you try there is no way you can add a fourth stick to the group that is perpendicular to the other three. This is, in essence, what we mean when we say that space is three-dimensional.

What are some of the properties of three-dimensional space?

First, we know that the shortest distance between two points in space is a straight line. We also know that two straight lines that start off parallel will remain parallel forever. It seems, in fact, that we know quite a bit about our three-dimensional space. But do we thoroughly understand it? The answer has to be no. As we saw earlier Newton was sure he understood it back in the 1700s. He thought space was absolute—in other words, the same for everyone throughout the universe. But Einstein proved that this isn't so. He showed that space could be bent and twisted, and that the amount of bending, or the curvature, can vary considerably from region to region.

Is it possible to see this curvature? No, we can't see it directly, but some people feel that they can visualize it to some degree. For most of us, though, visualization is impossible; we just don't have enough imagination. To see how we might try to imagine it, let's start with a sheet of paper. If you hold it up in front of you and bend it slightly, you see a two-dimensional curved surface. What's important here, though, is that the sheet is in three dimensions, and you need the extra dimension to see the curvature. This means that if we want to see curved three-dimensional space, we need an extra dimension, a fourth dimension.

There is a way, however, that can help us visualize three-dimensional curved space without introducing a fourth dimension. It's not a highly accurate representation by any means, but it is helpful. We can think of space as filled with jelly. Earlier we talked about space being filled with aether, and some people might have visualized this aether as a jelly. Einstein showed, of course, that we don't need the aether, but it's still convenient to think of space as filled with a sort of jelly. Then, when we talk about the bending and twisting of space, we can think of the bending and twisting of the jelly. At least it gives some idea of what is going on, and it's better than trying to visualize the bending of "nothing."

Most important, though, is that scientists have a way of describing the amount of bending. In other words, they need a mathematical theory of the bending of space, a geometry.

THE GEOMETRY OF SPACE

The first and simplest geometry is the one you learned in high school. It was devised by Euclid and his contemporaries, and was published in his book *Elements*. *Elements* is an amazing book, the first really rigorous book in mathematics, and the first that was based on logic. It has served as a model for mathematical theories ever since it was first published. Beginning with a number of fundamental definitions, basic postulates (something that is assumed), and five self-evident truths, or axioms, all of geometry is developed through the derivation of theorems.

For 2,000 years Euclidean geometry stood as the basic geometry of the world. It was logically and empirically consistent, and seemed to be the only consistent geometry that existed. In time, though, mathematicians began to consider the possibility that it contained a flaw. The problem was the fifth axiom. This axiom can be stated as follows: Through a point not on a given straight line only one line can be drawn parallel to the given straight line.

Was this really an axiom? Or could it be derived from the other four? Many mathematicians believed that it could be. There was no doubt that it is self-evident, but self-evidence is sometimes deceptive; you have to be careful. Many attempts were made to prove that it was superfluous, but no one succeeded. Eventually, however, mathematicians showed that this axiom was mathematically equivalent to the statement: The sum of the interior angles of a plane triangle is 180 degrees.

Certainly if you draw any type of triangle on a flat surface and measure the angles you find that they always add up to 180 degrees. But let's turn things around and ask: Is it possible under any conditions to have a triangle where the sum of the interior angles is not 180 degrees? The key words here are "under any conditions." If you think about it for a moment, you will realize that if the surface that you draw the triangle on is curved, the angles will not add up to 180 degrees. So we may indeed have another geometry.

GAUSS

One of the first to consider non-Euclidean geometries was Friedrich Carl Gauss of Germany. He spent many years trying to prove that the fifth axiom was superfluous, and, in the process, made many important contributions to mathematics. Strangely, though, he did not publish his discoveries, but he did communicate many of them to others, and they were published after his death.

Born in 1777 in Brunswick, Germany, Gauss was a child prodigy by anyone's standards. He could add and subtract by the time he was three and learned to read and write before he went to school. His father was a laborer—a gardener and part-time bricklayer—who did little to encourage his son's education. He was harsh almost to the point of brutality. Fortunately, his mother, a woman of strong character and sharp intelligence, and her brother Friedrich (who Gauss later called a genius) realized that Gauss had a gift—a prolific and quick mind—and they did everything within their power to help develop it.

Gauss entered school at the age of 7. It was a harsh school and caning was common. Little is known of his first two years at school, but when he was 10 he began his first class in arithmetic. One of the best known anecdotes of his youth happened soon after this, an event that changed his life. His teacher, J. G. Büttner, asked the students to add up all of the numbers between 1 and 100; he then sat back thinking he could relax for an hour or so.

Each of the students had a slate to work on, and when they were finished they were to write the answer on the slate and place it on Büttner's desk. In less than a minute, Gauss walked up to him. "Here it is," he said, placing the slate on his desk. He then sat back with his arms folded as the other students struggled with the assignment. Büttner looked at him suspiciously, certain there was no way he could have obtained the answer so quickly.

Finally, when all the slates were on the desk, Büttner looked them over. The only one to get the correct answer was Gauss. Puzzled, he asked Gauss how he had obtained it so quickly. Gauss

Friedrich Carl Gauss.

explained that he had looked at the sequence 1 to 100 and noticed that the first and the last added up to 101, the second and second to last also added to 101, and so on all the way to 50 and 51. There were 50 such pairs so he multiplied 50 by 101 and got 5050.

Büttner was amazed. He asked Gauss a number of questions about arithmetic and it was soon evident that the student knew more than the teacher. Büttner realized that there was nothing he could teach him, so he ordered an advanced book on arithmetic and gave it to him to study on his own. Within a short time, Gauss had mastered it.

Although Gauss worked mainly on his own, he did get some help from Büttner's assistant, Martin Bartels. Bartels was only eight years older than Gauss, but he was planning on becoming a mathematics teacher, and had studied mathematics on his own.

Bartels realized it was important for Gauss to get a university education, so he went to Gauss's father and talked to him. But Gauss's father was a stubborn man, and not easily persuaded. He had other plans for Gauss; he wanted him to learn a trade, such as bricklaying, and follow in his footsteps. Besides, he said, he didn't have enough money to pay for his son's university education. Bartels persisted, promising that a patron would be found to support him—and his persuasion finally paid off. Gauss's father grudgingly agreed to let him go to university on the condition that he didn't have to pay. Surprisingly, he even agreed to release Gauss from some of his chores so that he would have more time to spend on his studies.

A patron was, indeed, eventually found for Gauss. Bartels introduced Gauss to a professor Zimmerman at a nearby college, who, in turn, arranged for him to be brought before Duke Carl Wilhelm Ferdinand of Brunswick. Gauss was only 14 at the time, but his modesty, good manners, and obvious brillance impressed the Duke so much that he immediately agreed to sponsor him.

At 15 Gauss entered Caroline College in Brunswick, where he spent three years studying the works of Newton, Euler, and Lagrange. At 18 he went on to the University of Göttingen. The three years he spent at Göttingen were extremely important in his development; one of his first important discoveries was, in fact, made while he was at Göttingen. He showed how, using a straightedge and compass, you could construct a 17-sided polygon. This was a well-known unsolved problem of mathematics at the time, and Gauss solved it elegantly, much to the surprise of the mathematicians of Europe. He was so happy with his solution, he told a friend, Wolgang Bolyai, that he wanted a 17-sided polygon engraved on his tombstone. When he died engravers did try to put one on it, but decided that it looked too much like a circle and changed it to a 17-pointed star.

Bolyai was one of his closest friends at this time, and remained a lifelong friend. Like Gauss, he was also interested in the foundations of geometry. They spent many hours together talking about the problems. It was during this time, in fact, that Gauss developed his non-Euclidean geometry. Strangely though, he did not share his discovery with Bolyai until many years later.

Shortly after he left Göttingen, Bolyai became convinced he'd proven that the fifth axiom was superfluous, and he sent his proof to Gauss. Gauss looked it over and soon found a flaw in it. Knowing that Bolyai would be disappointed, Gauss worded his reply to him very carefully. And Bolyai was very disappointed, but he soon realized that Gauss was right.

As it became obvious that Gauss had done a lot of work in the area, friends—in particular, the mathematician Friedrich Bessel—encouraged him to publish them. But he refused, saying, "Perhaps it will not happen during my lifetime, since I fear the Boeotian cries if I were to express my opinions clearly." He is referring to the cries of second-rate mathematicians, and others who barely understood him.

The geometry that Gauss discovered is now called hyperbolic geometry, because it is the geometry of a hyperbolic surface. On this surface, the sum of the interior angles of a triangle is less than 180 degrees. Furthermore, more than one parallel line can be drawn through a point parallel to another line. Gauss later wondered if the space of our experience was, perhaps, non-Euclidean. He even went as far as performing an experiment to see if this was the case. Using the peaks of three nearby mountains he measured the interior angles of the triangle formed by them, trying to determine if their sum was exactly 180 degrees. The experiment was inconclusive.

Gauss spent most of his life at Göttingen. He married twice and had a total of five children. When he died, his works, taken mostly from his diaries, were published in 12 large volumes. He made important contributions to literally every branch of mathematics.

BOLYAI, LOBACHEVSKI, AND HYPERBOLIC GEOMETRY

Because most of Gauss's discoveries in geometry were not published until after his death, his non-Euclidean geometry was discovered independently by two other mathematicians. Both published their results, so the credit for the first non-Euclidean geometry generally goes jointly to them.

The first to discover the new geometry was Wolfgang Bolyai's son Johann (Janos). He learned about the problem from his father and became fascinated with it, but he didn't start working seriously on it until after he became an officer in the army. He wrote his father telling him of his work on the problem. Alarmed, Wolfgang wrote back telling him to leave it alone. "It will become a poison all your life," he said. And indeed he was speaking from experience; he had worked on it for years without any real progress.

Finally, in 1823 Johann wrote his father telling him that he had used the fifth axiom to construct a new geometry. He enclosed a copy of his work with the letter. Wolfgang looked it over, but didn't know what to make of it, so he sent it to Gauss for his comments.

Gauss sent it back saying, "I believe that the young geometer Johann Bolyai is a genius of first rank." But he went on to say that although he was convinced that the geometry was correct he wouldn't be able to publicize it. He said he had discovered the same geometry almost 30 years earlier and had left it unpublished. His diary, found after his death, confirmed that he had.

Wolfgang was satisfied with the reply, but his son was not. He was annoyed that Gauss claimed to have discovered the geometry earlier, and therefore would do nothing to help him. Nevertheless, Bolyai's article was soon published as an appendix in one of his father's books. But for many years it went unnoticed.

Johann became bitter over the experience. A few words from Gauss would have made a significant difference. Soon after this he retired from the army and lived out the rest of his life on the family estate, but he did no further work in mathematics.

In the early 1800s, a Russian mathematician, Nicholas Lobachevski, also began looking at the fifth axiom. As Bolyai had earlier, he also soon found that a non-Euclidean geometry was possible. He developed it and wrote several papers discussing it.

Lobachevski was born in 1793. His father died when he was only a few years old and he moved with his mother to Kazan. He started school in Kazan at age 8 and made rapid progress. By the time he was 14 he was ready for university.

He graduated from the University of Kazan with a masters degree when he was 18. Two years later he obtained a Ph.D. and soon became an assistant professor at the university. His duties were heavy; he taught mathematics, physics, and astronomy in addition to doing research. Within a few years he was also head of the library and museum.

Finally, in 1827, at the age of 34 he became Rector (President) of the university. Despite his position he continued to work in the library and museum. He could frequently be found in old clothes dusting and cleaning. One day he was working in the library when a distinguished visitor from another country came in. Thinking he was the custodian, or perhaps the janitor, the visitor asked to be shown around. Lobachevski gave him a tour, explaining things in detail as they went from room to room. The visitor was so impressed with his knowledge he offered him a tip as he left. Puzzled and insulted, Lobachevski refused it. The visitor returned the money to his pocket, shrugged, and left. That evening they met again at the Governor's table—much to the surprise of the visitor.

Lobachevski worked on his geometry for over 20 years. He announced his discovery in 1826 at a meeting of the Physics and Mathematical Society of Kazan. Gauss, however, didn't hear about it until 1840. As soon as it was brought to his attention, though, he wrote Lobachevski praising the work; he even arranged to make him a member, in absentia, of the Göttingen Academy.

Nicholas Lobachevski.

As mentioned earlier, Lobachevski's geometry was the geometry of a negatively curved or hyperbolic surface. It is easiest to visualize this surface as like that of a saddle. The sum of the interior angles of a triangle on such a surface is less than 180 degrees.

Lobachevski eventually published a book on his new geometry and a copy of it soon found its way into the hands of Johann Bolyai. Looking through it, Bolyai became even more embittered, realizing that it was the same geometry he had discovered years earlier. He was convinced that Gauss had conspired to deprive him of the honor of the discovery. But Lobachevski had discovered it independently of both Gauss and Bolyai.

RIEMANN AND ELLIPTICAL GEOMETRY

As it turned out the Bolyai–Lobachevski geometry was not the only non-Euclidean one. It is perhaps strange that both men, and Gauss, discovered only one type of non-Euclidean geometry, for there is a second type that now seems quite obvious to us (of course hindsight is always much easier than foresight). In the two-dimensional geometry of Bolyai and Lobachevski the sum of the interior angles of a triangle is less than 180 degrees. An obvious extension of this is to a positively curved surface where the sum of the interior angles is greater than 180 degrees. None of the three men made this extension; this discovery was left to Bernhard Riemann.

Born in Hanover, Germany in 1826, Riemann, like Gauss, was a child prodigy. His family was poor, but from all indications he spent a happy youth. His father was a Lutheran minister who hoped that Bernhard would follow in his footsteps.

Riemann's mathematical talents became obvious soon after he entered school at the age of six. He was easily able to solve any problem that was put before him. In fact, he enjoyed making up difficult problems that others were unable to solve. When he was 14 he entered the gymnasium. He was extremely shy and was teased continually, but this treatment had little effect on his schoolwork, which was uniformly excellent. Two years later, he transferred to a different gymnasium where the director noticed his talent in mathematics and gave him free reign of the library. He was soon studying calculus and other difficult branches of math.

At 19 he entered the University of Göttingen. He was still planning on becoming a minister, but soon realized that his real love was mathematics. He was reluctant to tell his father but finally confessed to him that he wasn't interested in the ministry and

Bernhard Riemann.

wanted to pursue a career in mathematics. His father was disappointed, but agreed that it was best under the circumstances.

In 1851, Riemann obtained his Ph.D. He now wanted to become a privatdozent—the first step in the path to a professorship—but first he had to convince the faculty at Göttingen of his qualifications. This meant that he had to present a lecture to a committee of three professors. If they were satisfied, he would become a lecturer, but his only pay would come from the students in his classes.

Riemann presented three topics to his committee. He was

only well-prepared to lecture on two of them, but he was sure that one of the two would be selected. The third talk he listed was on the foundations of geometry. One of the committee members, however, was Gauss, and when he saw the third topic it brought to mind his own early work on the subject. The topic intrigued him and he wondered what Riemann would do with it. He therefore selected it.

Riemann was shaken up when he heard the news. He hadn't expected his third topic to be selected. To make things even worse he was heavily involved with research at the time and didn't want to spend a lot of time preparing a new topic. He was in a quandary, but he knew that he had to do well. So he put everything else aside and dug into the subject with vigor, and in the process made several important discoveries. He titled his lecture "On the Hypotheses Which Lie at the Foundations of Geometry." The lecture was a resounding success; Gauss was astounded. Only occasionally did he praise students, but in Riemann's case he couldn't praise him enough.

Needless to say, Riemann passed the exam, and was soon lecturing for a fee. He expected his first class to be small, perhaps two or three students, and was pleased when eight showed up. A few years later he was given a permanent position with a small salary, then, in 1859, he was selected as Gauss's successor.

He married at the age of 36, but his life from that point on was anything but happy. About a month after he married, he fell seriously ill and had to quit teaching. He went to Italy to recuperate, but it helped only briefly. His sickness was soon diagnosed as consumption. For the next three years he taught occasionally when he felt strong enough, but he gradually grew weaker and weaker until finally in 1866, at the age of 39, he died.

The non-Euclidean geometry that Riemann discovered was that of a positively curved surface (the surface of a sphere), or more generally, an ellipsoid (an egg-shaped surface). It is therefore called elliptical geometry. If you draw a triangle on a sphere, it is easy to see that the sum of its interior angles is greater than 180 degrees. Furthermore, it's easy to see that no line can be drawn

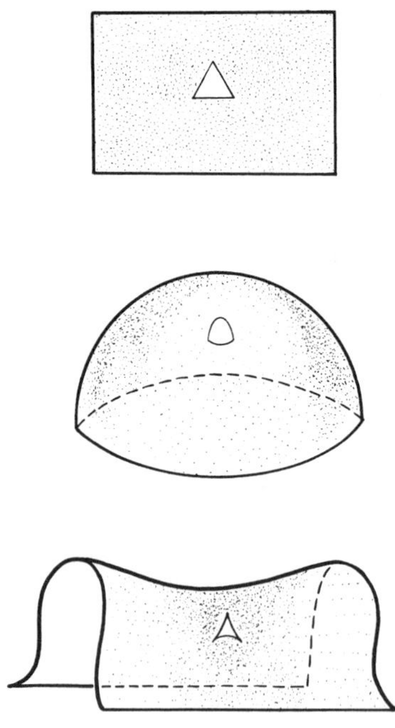

The three geometries in two dimensions: flat (top), positively curved (center), and negatively curved (bottom).

through a point parallel to another straight line. The best way to prove this for yourself is to look at a globe of the earth. The lines of longitude start out parallel at the equator, but they all meet at the poles.

Of particular importance, though, is that Riemann did not think of his geometry as necessarily that of a surface. He visualized the extension to three, four, and even more dimensions. Furthermore, he considered the possibility that the curvature of the space might vary from point to point. We refer to this as a local geometry, in contrast to a global geometry where the entire space is curved in the same way.

EXTENSION TO THE FOURTH DIMENSION

Two geometries, besides Euclidean geometry, had now been formulated: one associated with a positively curved surface (in two dimensions) and one with a negatively curved surface (in two dimensions). But if these geometries were to describe curved space, they had to be extended to three dimensions. How can we do this? Mathematically, it's relatively simple, but unfortunately, it's difficult to visualize three-dimensional curved space. In fact, it may be impossible, but we'll try anyway.

The best way to begin is by analogy. As we saw earlier we can easily visualize a curved two-dimensional surface in three-dimensional space. Just take a sheet of paper, hold it up in front of you, and bend it. You are looking at a curved two-dimensional surface. To see the curvature, however, you need a three-dimensional space around it. This means that if we are to see curved three-dimensional space, we need an additional dimension around it—a fourth dimension.

But this presents a problem. Our experiences, our senses, and so on are all geared to three dimensions. It is therefore difficult for us to relate to four dimensions. Let's look at how we might visualize it, however. Again, we'll work by analogy. Starting with a point of zero dimensions, we move it sideways and get a one-dimensional line. In the same way, if we move the line perpendicular to itself, we get a two-dimensional surface. And, finally, if we move the surface perpendicular to itself, we get a three-dimensional square.

What would we get if we continued this to four dimensions? We would again have to move the square perpendicular to itself. It's hard to see how we could do this; one way would be to just move it sideways. Or perhaps it would be better to expand it outward. Both of these are shown below. We're not sure which gives a better representation of four dimensions, but at least they give us some idea of what a cube might look like in four-dimensional space.

It turns out that there's something else we can say about four-

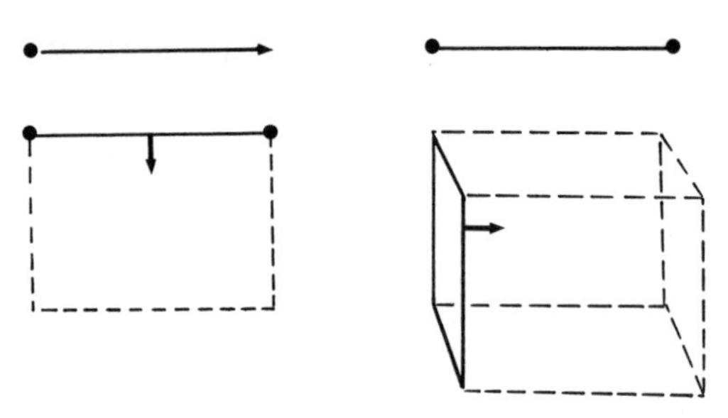

A progression to higher dimensions. If you move a point, you get a one-dimensional line. If you move a line you get a two-dimensional surface. If you move a surface, you get a three-dimensional volume.

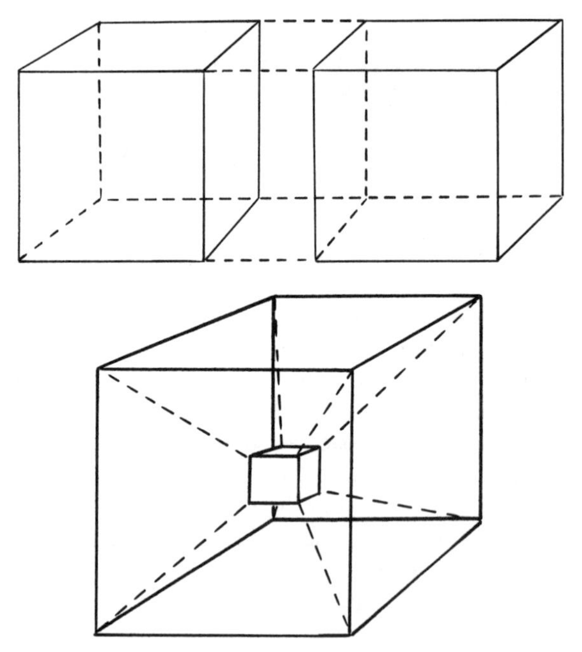

Two representations of what four-dimensional space might be like.

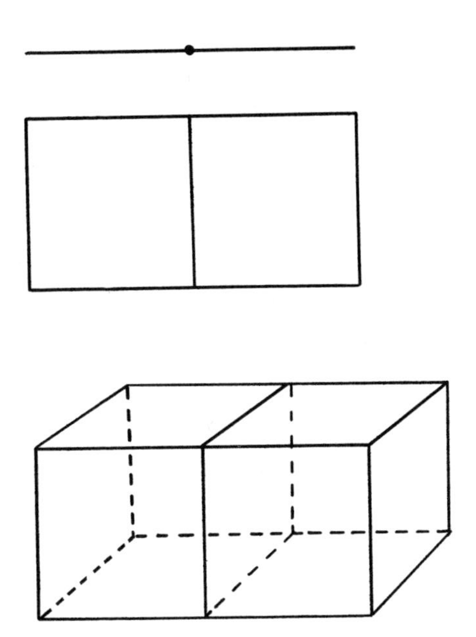

A point splits a one-dimensional line into two parts. Similarly, a line splits a two-dimensional surface in two parts. Finally, a surface splits a three-dimensional column in two.

dimensional space. Our three-dimensional space must somehow split it in two parts. To see why, consider the following. If we place a zero-dimensional point anywhere on a line, it splits it in two parts. Similarly, if we draw a one-dimensional line across a two-dimensional sheet it splits it in two pieces. Finally, a two-dimensional sheet that extends across a three-dimensional box splits it in two parts. It is therefore reasonable that our space of three dimensions would split a four-dimensional space into two sections. But, again, how we should visualize this is difficult to say.

Many people prefer to call four-dimensional space, hyperspace. And it is perhaps an appropriate name. We certainly can't experience hyperspace directly; to us it's nonspace. But, as we will see, it may play an important role in the universe. In particular, if we are to visualize three-dimensional curved space, we need hyperspace around it.

HYPERSPACE TIME TRAVEL

Let's go back again to our two-dimensional sheet in three-dimensional space. Two-dimensional beings, assuming such things existed, could only travel from point to point along the sheet. Even if the sheet was bent, they would still have to move along the surface to get from some point A to another point B. We can see, however, that there is a shortcut. If the two-dimensional beings knew about three-dimensional space, they could cut across it, and in the process they would save themselves considerable time. Note that we can apply this to both positively curved surfaces (Riemann) and negatively curved ones (Lobachevski–Bolyai).

Now let's extend this to three-dimensional space. We would have to assume our curved three-dimensional space is surrounded by hyperspace. We can draw a simple representation of it as below. Again, as in the above case, we see a shortcut. This time it is through hyperspace. But we have the same problem as the flat two-dimensional beings do. They are unfamiliar with three-dimensional space and don't know how to take advantage of it. They have no idea how to get from two-dimensional space to three-dimensional space. And, of course, we have no idea how to get to hyperspace. If we did, though, we could travel between A and B in a shorter time than it would take in three-dimensional space. At first glance, however, it appears as if it isn't much shorter. In traveling to the stars, we're obviously going to need a much bigger advantage than this shortcut through hyperspace appears to give us.

It turns out, though, that we may have an additional advantage. We are assuming that time passes in the same way in

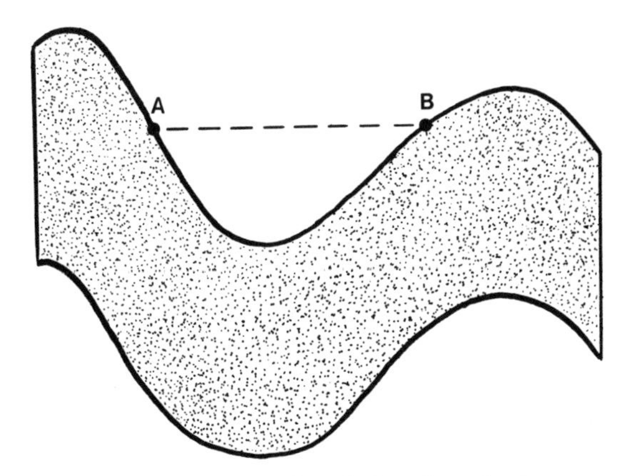

A shortcut from A to B through three-dimensional space.

hyperspace as it does in the three-dimensional space of our experience, and this may not be true. Several scientists have speculated that time may not exist in hyperspace. Or perhaps I should say that it would not pass. If this were the case, the trip from A to B would be extremely short, as no time would pass while we were in hyperspace. It's important to remember, though, that this is pure speculation. It may or may not be true, but if it is, it would certainly be helpful.

Let's turn now to the exact relationship between three-dimensional space and hyperspace. Again, we can only speculate. A number of people have suggested that if we were capable of seeing hyperspace, we would see a Swiss cheese world. Throughout hyperspace we would see branches and loops of space. If this is the case, we should be able to take advantage of hyperspace to travel between two points in space.

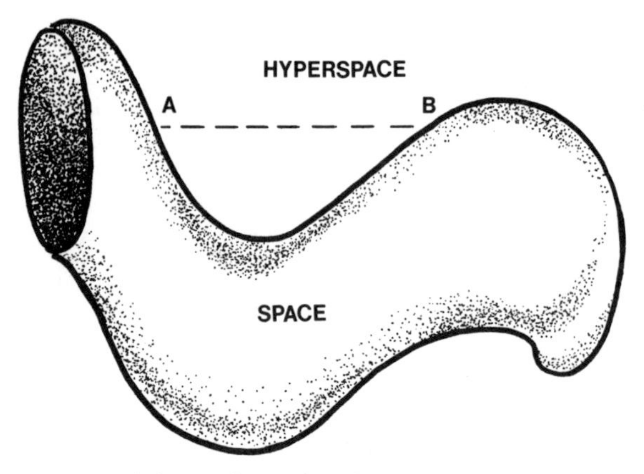

A shortcut from A to B through hyperspace.

GLOBAL AND LOCAL CURVED SPACE

In practice, there are two levels at which we can think of curved space. In cosmology, which is the study of the structure of the universe, astronomers worry about the overall, or global curvature of space. According to current theories, the universe can be positively curved, negatively curved, or flat. We're still not sure which is the case. This is, in fact, one of the central problems of cosmology: determining the overall shape of the universe. Its shape is related to its future. We know, for example, that the universe is expanding; in other words, all galaxies, or, more exactly, all groups of galaxies, are moving away from all other groups of galaxies. If the universe is positively curved, its expansion will eventually stop and it will collapse back on itself. If it is negatively curved, on the other hand, it will continue expanding forever.

Of more interest to us, however, is local curvature, or curvature on a small scale compared to the overall universe. A passage through hyperspace will obviously have to be a region of extreme curvature—a sort of "tunnel"—and therefore it will be small.

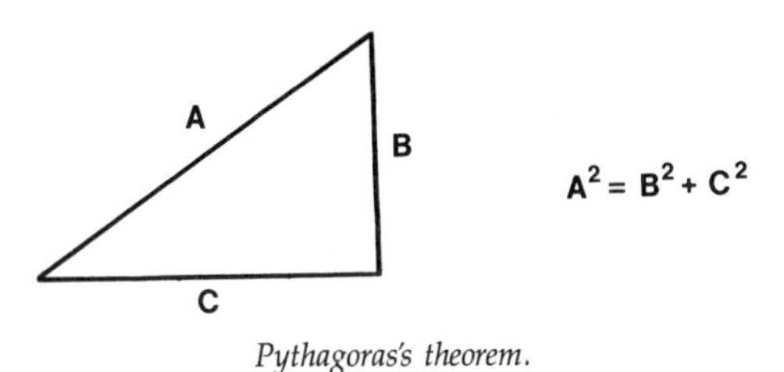

Pythagoras's theorem.

Riemann was the first to consider curved space on a local scale. He realized that he could use Pythagoras's theorem to measure the curvature at any point in space. You probably remember Pythagoras's theorem from your high school geometry. It states that for a right-angle triangle, the square of the hypotenuse (sloped side) is equal to the sum of the squares of the other two sides.

Consider how this theorem would change if we applied it to a curved surface. If we draw small triangles on a curved surface and measure the sum of the interior angles, we find that in some cases we get more than 180 degrees and in others less. The difference from 180 degrees depends on how curved the surface is. The deviation from Pythagoras's theorem therefore gives us a measure of the amount of curvature and how it varies from place to place. Again, we would like to extend this to three dimensions, and Riemann showed us how to do this.

Riemann's ideas on local curved space were championed and extended by the English mathematician William Clifford. In 1870 he proposed that matter was made up of curved space and that the motion of matter was associated with a wave of curved space. In his paper, "On the Space Theory of Matter," he made the following four hypotheses.

1. Small regions of space are, in fact, of a nature analogous to little hills on a surface which is, on the average, flat; namely, that ordinary laws of geometry are not valid in them.

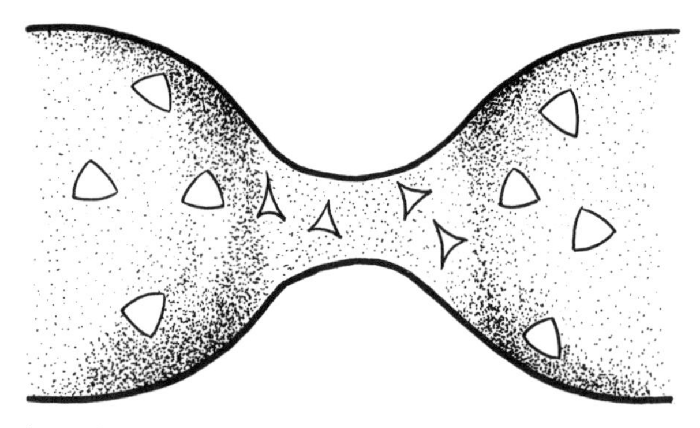

A surface with varying (local) curvature. Small triangles show type of curvature.

2. This property of being curved or distorted is continually being passed on from one portion of space to another after the matter of a wave.
3. The variation of the curvature of space is what really happens in the phenomena which we call the motion of matter.
4. In the physical world nothing else takes place but this variation.

OTHER UNIVERSES

So far we have only discussed the possibility of travel through hyperspace to some other point in our universe. It is also possible that if we did take such a trip, we could end up in a different universe. It's difficult to comprehend how another universe might differ from ours, or even how we would be able to distinguish it from ours. Nevertheless, it is an idea that has been taken seriously and is talked about by the scientific community.

To see how we might get to another universe, let's begin with how we might travel to a distant point in our own universe. If we

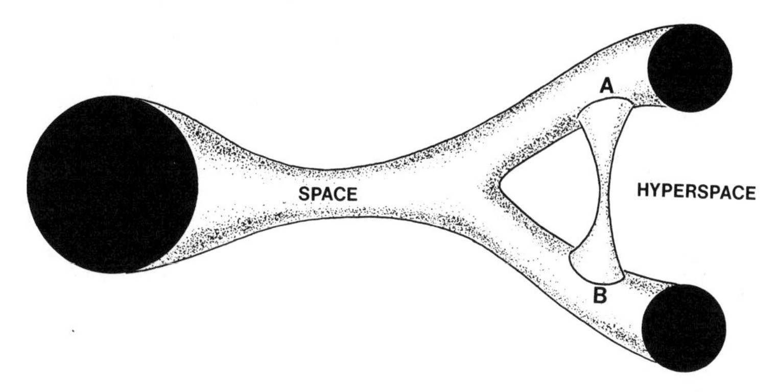

A "tunnel" through hyperspace to another region of space.

were to travel between points A and B in the figure above , we would need a "tunnel." We can easily visualize what this tunnel might look like.

It is possible, however, that we won't end up in our own universe when we pass through such a tunnel. To understand

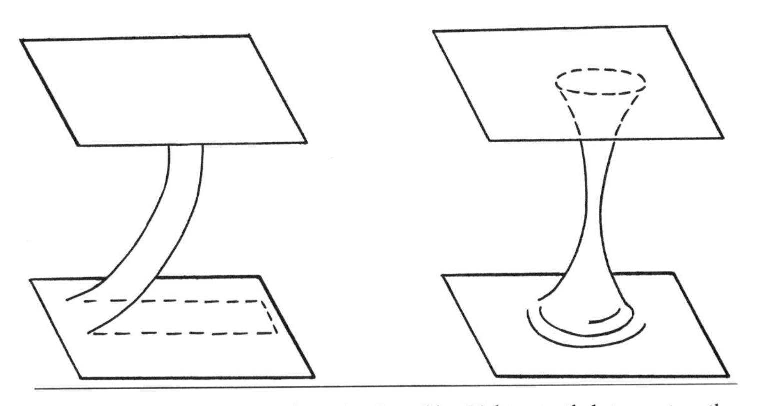

Left: a path between two two-dimensional worlds. Right: a path between two three-dimensional worlds.

why this might be the case, let's go back again to two dimensions. Assume that we have two two-dimensional worlds, i.e., two sheets. How would the beings in one of the worlds pass to the other? Obviously if we cut one of the sheets, bend it up, and join it to the second we would have a route. For simplicity let's show this route as a curved surface. If this surface connected the two two-dimensional worlds, beings from one of them could move along it and pass from one world to the other. In the case of three-dimensions there would be a three-dimensional universe at either end of the tunnel, and the beings would use the interior of the tunnel. This means that if we had two entirely separate universes sitting side-by-side, separated by hyperspace, it might be possible to use such a tunnel to pass from one universe to the other.

The real problem, however, is: How could we generate such a tunnel? Or, more generally, how could space become curved? Obviously something has to curve it. We will see in the next chapter that it was Einstein who provided the answer to this.

CHAPTER 4

Taming the Curvature

We have seen that if space *is* curved, we may someday be able to travel rapidly and easily through it to distant points in the universe. So far, though, we have said nothing about how space becomes curved, or if, indeed, it happens at all. What would cause space to curve?

Although people such as Riemann and Clifford talked about the possibility of curved space, the first person to show that it actually could become curved was Einstein. Using intricate mathematics he demonstrated that matter caused space, or, more exactly, space-time, to curve. The space-time around our sun, for example, is curved because of the mass of the sun. We see evidence of this curvature in the way the planets move; they follow the curvature.

How did Einstein arrive at this conclusion? It certainly wasn't an idea he arrived at overnight; it took ten years of intense and sometimes frustrating work to prove it. After completing his special theory of relativity, as expected, he tried to extend to nonuniform, or accelerated motion. In attempting to do this, though, he encountered a serious difficulty. Unlike uniform motion, accelerated motion appeared to be absolute; in other words, you could detect it without reference to another system. You feel a force on your body when you are accelerated or decelerated. To illustrate further, assume that you are on a silent, smoothly running train. If you look out the window and see another train moving past you on a nearby track, you can't be sure whether you are moving past it or if you are sitting still and it is moving past you. But, if the brakeman suddenly puts on the brakes, you know you're moving. Unless you hold on, you will be thrown into the seat in front of you.

Accelerated motion, it seems, is absolute. But Einstein knew that if he was to generalize special relativity to include acceleration as well as uniform motion, the two types of motion had to be put on the same footing. Acceleration, therefore, couldn't be absolute, and there had to be some way to prove it wasn't. But how?

Einstein was sitting at his desk in the patent office one day when the solution to the problem came to him. Gravitation was the key. When a person falls or jumps from a high building, he accelerates, but he does not feel his weight; there is, in effect, no gravitational field acting on him. This was the missing link, the magic ingredient he needed. He later referred to it as the "happiest thought of his life."

To see why Einstein was so excited, let's begin with Newton's law of inertia. It states: The force needed to accelerate a body to a given speed depends on the mass of the body. Turning this around, we see that we can use it to determine the mass of an object. Just give the object a push with a known force and see how fast it accelerates. The mass you obtain in this case is called the inertial mass. But Newton was also able to determine the mass of a body using his law of gravity. According to this law, all bodies in the universe attract one another, so all you have to do is measure how fast the Earth pulls a falling body to its surface. The mass obtained in this case is called the gravitational mass. From a numerical standpoint, gravitational and inertial mass are the same, but Newton never thought of this as anything special; to him it was just a coincidence of nature.

To illustrate the difference more fully, suppose you have several boxes ranging from small light ones up to large heavy ones. If you try to slide them across a well-waxed floor, you know that it takes much more force to push the massive ones than the lighter ones. Also it takes a lot more force to stop the massive ones once they are in motion. In fact, with a little experimentation, you can show that the greater the mass, the greater the force you need for the same acceleration. The mass we are dealing with here is the inertial mass.

Now assume that you drop the boxes from a window—all at the same time. You see immediately that they all fall to the ground with the same acceleration, and therefore reach it at the same time. But the larger boxes are much more massive than the smaller ones, and the gravitational pull on them is therefore greater. In fact, if you double the mass of an object, the earth pulls on it with double the force. Yet strangely the massive boxes still fall to the earth at the same rate as the light boxes. Why? Obviously there is something holding them back—something stopping them from accelerating faster than the lighter ones. And, of course, it's their inertia; as we just saw, it takes more force to move a massive box than a light one. The extra force needed to get the more massive box moving is therefore balanced by the extra force that is exerted on it by the earth. What we are really saying here is that objects fall at the same rate because inertial mass (or force) is equal to gravitational mass (or force).

For many years there was a general acceptance of the equivalence of these two types of mass. No one challenged the idea, but few took the time to wonder why it was true. There was, however, considerable curiosity about the equivalence, and whether it really was exact. In 1889, Roland von Eötvös of Hungary showed that it was; he demonstrated that the two masses were equal to one part in a hundred thousand. In 1964, R. H. Dicke of Princeton showed that they are equivalent to one part in a hundred billion.

There's another way of looking at this equivalence, one that Einstein found quite useful. To illustrate, we'll use what he referred to as a "gedanken" experiment—an experiment we perform in our mind. Einstein used such experiments extensively in deriving his general theory of relativity. We'll begin by assuming that you are in a small, windowless room on Earth. When you drop an object it will accelerate downward at the acceleration of gravity—in other words, it will drop 32 feet per second squared. Assume now that you go to sleep and, while you're sleeping, the room is transported into space and accelerated upward at a rate equal to the acceleration of gravity. When you wake up, you will find that there is a force pushing you toward the floor of the room, just as there is on Earth; in fact, you will feel exactly the same. This means that if you dropped a ball, it will accelerate toward the floor in exactly the same way it does on Earth. You will therefore have no idea that you are in space; it will seem as if you are still on the surface of Earth. Indeed, no matter how many experiments you performed, you will not be able to detect a difference.

Einstein referred to this as the principle of equivalence. It can be looked upon as either the equivalence of the room on Earth and the accelerating room, or the equivalence of gravitational and inertial masses. They are just two different ways of looking at the same principle.

The equivalence of inertia and gravity is well known to pilots. When flying through fog or clouds, they have to be careful when banking their plane steeply, as it is difficult to distinguish forces due to gravity from those due to inertia. As strange as it may seem, it is possible for a pilot to lose his sense of direction relative to the earth.

If you have been observant, though, you may have noticed a slight flaw in the principle of equivalence as it is stated above. It is possible to distinguish the "gravitational field" in an accelerating room from the gravitational field of, say, the Earth. As you move upward in the earth's gravitational field, it decreases in strength, so there is a slight difference in gravitational pull on your feet as compared to your head when you are standing. The pull on your feet is stronger. The "field" of an accelerating room, on the other hand, is uniform. To get around this difficulty we say that the accelerating room is extremely small. Actually, for a strict equivalence, it would have to be infinitely small, but we'll just assume it's small enough so that there is not a significant change over its height.

I mentioned earlier that Newton knew about the equivalence of inertial and gravitational forces. If so, what exactly was Einstein's contribution? Prior to Einstein, scientists assumed that no mechanical experiment would ever allow them to distinguish

The equivalence principle. An apple in an elevator that is accelerated upward with an acceleration g (acceleration of gravity on Earth) will drop in the same way an apple does on Earth. The force of inertia (left) and the force of gravity (right) are equal.

between inertial and gravitational forces. They felt, however, that there really was a difference and it would eventually show up in optical and electromagnetic experiments. Einstein disagreed. He said that in accelerating the elevator upward we weren't creating an "artificial" gravitational field, we were creating a "true" gravitational field. There was no difference whatsoever between the two cases and no experiment of any kind would ever distinguish them.

Einstein was pleased with his principle of equivalence. It allowed him to make several important predictions. The first was that a beam of light would be bent by a gravitational field. To see how he arrived at this conclusion, let's go back to our small room in space. Assume that this room is moving at constant speed, and that a beam of light is allowed to enter through a small window in its side. Mark the spot where the beam hits the wall on the opposite side. Then accelerate the elevator upward and watch the beam; as you might expect it will be deflected downward. But an upward acceleration is equivalent to a gravitational field via the equivalence principle, and this means that a beam of light will be deflected by a gravitational field.

In a similar thought experiment, Einstein showed that a beam of light coming to us from any gravitating object, the sun for example, would have its vibrational rate, or frequency, decreased. This is usually referred to as the Einstein redshift. What is particularly important about it is that it implies that the rate of a clock would also be affected. It tells us that a clock in a strong gravitational field will run slower than one in a weak field.

All of these results were summarized in a paper that Einstein published in 1907. It was actually a review paper on special relativity, but toward the end of the paper he talked about how the theory could be extended. The key to extending it to nonuniform motion, he said, was gravity and the equivalence principle. He was pleased with the result because it was, in a sense, killing two birds with one stone. A generalization of special relativity, he realized, would cover all natural phenomena except gravity. But with the principle of equivalence, gravity would be included.

At this stage in his career, Einstein was still working at the patent office, and, naturally, he wanted something more—in particular, a university position. But, to become a university professor, he first had to become a privatdozent—a lecturer who is paid only by the students in his class. So he applied to the University of Bern, submitting his paper on special relativity as evidence of his scholarship and research activity. To his surprise his application was rejected; officials at the university said it was incomplete and that his paper on relativity was incomprehensible.

Einstein was dejected and decided to forget about an academic career. But professor Alfred Kleiner of the University of Zurich realized that Einstein was rapidly becoming an important figure in theoretical physics. He wanted him to come to the University of Zurich, so he visited him in Bern and encouraged him to reapply for a privatdozent. Einstein did, and, early in 1908, he was accepted. He was now able to lecture, but his life was no easier; in fact, in many ways it became more complicated. He still had to work during the day at the patent office, and now he had to teach in the evenings and on weekends. Furthermore, his classes were small, and he received little in the way of remuneration from them. In short, he was still poor, overworked, and now had even less time for research than before. To make things worse, Kleiner visited one of his classes and wasn't impressed with his teaching style. "I don't demand to be appointed a professor at Zurich," Einstein shot back when Kleiner brought it up.

ZURICH

Despite Kleiner's initial negative impression of Einstein's teaching ability, he still wanted him at the University of Zurich. And, in the spring, of 1909 a teaching position became available. Two names were put forward for it: Einstein and a colleague from his Zurich Polytechnic days, Friedrich Adler. When Adler heard that they were considering Einstein, he immediately withdrew his name, telling the authorities that Einstein's accomplishments far surpassed his own, and if they could get him, they should. Einstein was therefore hired as an associate professor in May 1909, effective the following autumn.

It's interesting to note that even before Einstein began his academic career he was awarded an honorary doctorate. Shortly after he had accepted the position at Zurich, while he was still working at the patent office, he received a large envelope from the University of Geneva. He glanced through it quickly, and thinking that it was of no consequence, he threw it into the wastebasket. The letter was an announcement that he, Marie Curie, and Friedrich Ostwald would receive honorary degrees at the summer graduation ceremonies. When authorities at the University of Geneva received no reply from Einstein they got in touch with him. He was pleased—and embarrassed when he realized that he had thrown the invitation away.

In autumn, with the beginning of the school year at Zurich, Einstein finally left his job at the patent office. He had worked there for a total of seven years; when he left he looked back on those years as some of the most productive of his life. Writing to Besso, he referred to the patent office as "that secular cluster where I hatched my most beautiful ideas." Einstein was now 30 years old. He hoped that now he'd finally have more time to devote to research. But this didn't turn out to be the case. Teaching and other duties took up most of his time; furthermore, he found teaching slightly frustrating. He loved to talk to students about physics; in particular he thoroughly enjoyed sharing his work with others, but at the same time he abhored the tremendous amount of time that routine classes took away from his research. He felt that each day was important, each day his powers of reasoning were weakening—and he had to take advantage of them while he was still young.

Yet, from all accounts, he was a good teacher. He explained things in detail, took time to answer questions, and usually gave numerous examples. Furthermore, unlike most professors, he was friendly with his students, sometimes taking them to cafés after class. And perhaps most importantly, he had an excellent sense of humor.

Einstein stayed at the University of Zurich from October 1909 to March 1911. During this time he published 11 papers, yet none of them were on his extension of special relativity. There is little doubt, however, that he was thinking about the theory.

Einstein seemed to enjoy Zurich. He was still among the poorly paid and there was always the pressure of lectures. Nevertheless, he seemed happy. Officials were therefore surprised when he suddenly announced that he was leaving for Prague.

TO PRAGUE

Einstein first heard that he was being considered by the German University in Prague in early 1911. If he obtained the position, he would be promoted to full professor and would be given a substantial raise. Understandably, he was anxious to obtain it. Two names were put forward: Einstein and Gustav Jaumann, a professor at the Technical Institute of Brno. Several of the officials preferred Jaumann, but according to custom the men had to be listed according to their accomplishments, and Einstein's were considerably greater than Jaumann's. Therefore Einstein's name was given precedence. Jaumann was insulted at being placed second and withdrew, leaving the position to Einstein.

Einstein and his family arrived in Prague in March 1910. Their standard of living improved with the move; they now had a live-in maid. "I'm having a good time here, even though it is not as pleasant as Switzerland," Einstein wrote soon after he got settled. And, within a short period of time, he returned to his work on the generalization of special relativity. One of the first things he did was calculate the bending that would be expected in a beam of light from a distant star as it grazed the sun. He found that the angular deflection would be 0.83 second of arc. In a paper published shortly thereafter, he encouraged astronomers to try to measure it.

Unknown to Einstein, the bending of light had been predicted in 1801 by a German astronomer, Johann Georg von Soldner. This was a slight embarrassment to Einstein in 1921 when the physicist Philip Lenard, an ardent Nazi supporter, pointed it out. Lenard, in fact, tried to use it to prove that Einstein stole most of his ideas, but Max Von Laue of the University of Berlin quickly set the record straight.

Einstein published two papers on his extension of relativity while in Prague. In them, he summarized what had been done on the principle of equivalence, the bending of light, and the redshift of light from the sun. It was during his stay at Prague that he began to see that there were going to be serious problems in extending special relativity. It appeared as if he was going to have to abandon Euclidean geometry. One thing that bothered him was the problem of the "rotating disk." According to special relativity, a rapidly rotating rigid disk would shrink along its circumference, yet its radius would remain constant. Near the speed of light, in fact, it would have a circumference near zero, but the same radius.

This was, of course, impossible—at least within Euclidean geometry.

Furthermore, there was something else that wasn't consistent with Euclidean geometry. A beam of light was usually considered to traverse a region of space along the shortest possible path—a straight line. Yet, according to Einstein's early calculations, a beam that grazed the sun followed a curved path. Euclidean geometry obviously didn't apply to light beams near the sun.

In both of these cases, it seemed as if a different, or non-Euclidean geometry would be needed, but Einstein had no idea how to proceed. During his stay at Prague he became friendly with a mathematician by the name of Georg Pick. He told Pick about his problem and Pick suggested that a new mathematics called "tensor theory" might be helpful. But there is no indication that Einstein looked into this possibility at the time.

In October of 1911, Einstein attended the first Solvay Congress, sponsored and financed by the wealthy Belgian industrialist, Ernest Solvay. Here he met for the first time many of the best known physicists of Europe: Lorentz, Rutherford, Curie, Nernst, and others. For a period of several days they discussed the basic problems of physics.

While Einstein was at Prague, his college friend Marcel Grossman, who was now at Zurich Polytechnic, was steadily rising through the ranks. Finally, he became dean of mathematics and physics, and, as dean, decided to get in touch with Einstein and see if he was interested in a position at the Polytechnic. As it turned out, Einstein was getting tired of Prague. He gladly accepted Grossman's offer, and in February of 1912 he returned to Zurich, this time, to his old alma mater, the larger and more prestigious, Zurich Polytechnic.

COLLABORATION WITH GROSSMAN

Einstein was now convinced that his new theory was going to be significantly different from his old one. In particular, the geometry had to be changed. He also decided that it was important that the mathematical form of the equations be the same in all coordinate systems. In other words, they couldn't vary from system to system. This is referred to as covariance.

Soon after arriving at Zurich Polytechnic, Einstein got together with Grossman. "Help me . . . or I'll go crazy," he is reported to have said to Grossman. He explained the problem to him and asked if there was any mathematics that could handle it. Although Grossman had not worked in the area of geometry for several years, he had done his doctoral thesis on non-Euclidean geometry. He looked into the matter and told Einstein that Riemann had developed a geometry that might be useful. Grossman and Einstein had both, in fact, taken a course several years earlier on Gauss's theory of surfaces. Riemann's work was closely associated with it.

Grossman agreed to help with the mathematics, but he didn't feel qualified to work on the physical aspects of the theory and wanted no part of it. He also informed Einstein that the new mathematical theory they would have to use, called tensor theory, was incredibly complex. He warned him that using it to generalize special relativity would be extremely difficult. And, as it turned out, it was even more difficult than Einstein anticipated. "Every step is devilishly difficult. . . . I have never been so tormented. A great respect for mathematics has been instilled in me. Compared with this problem the original relativity was child's play," he wrote shortly after beginning work with Grossman.

Grossman and Einstein published two papers together, and although they came extremely close to solving the problem, by a strange oversight, they missed it. They actually wrote down the proper equations at one point and examined them, but decided they were inadequate.

But, if the final equations were extremely complex, you might wonder how they could ever hope to derive them. It turns out that they had several things to guide them. First, when the gravitational field was extremely weak, the equations had to reduce to the well-known equations of Newton. In other words, they had to

Marcel Grossman.

be the same as Newton's equations. Furthermore, there were conservation laws, such as the conservation of energy, that the equations had to satisfy. Einstein also preferred that they be covariant, in other words, the same in different coordinate systems. In addition, they had to predict the bending of light and the Einstein redshift, and finally there was a well-known anomaly in Mercury's orbit that had never been explained. Einstein hoped the new equations would explain this anomaly.

For weeks, then months, the two men struggled, trying many different equations, and working out all of their ramifications. But nothing seemed to work. The only way, it seemed, to get a satisfactory set of equations was to abandon the condition of covariance. Einstein did so reluctantly, but, with it out of the way, they finally arrived at a set of equations that seemed satisfactory. Einstein was not entirely satisfied with the theory at this point, however; it appeared to be lacking in several respects. "It's marred by an ugly dark spot," he wrote to H. A. Lorentz in Leiden. The ugly dark spot he referred to was the fact that it was not covariant. He struggled with the problem for weeks, then finally convinced himself that covariance didn't matter. In fact, he derived two proofs that the theory need not be covariant. But there were other problems: the theory didn't account for the anomaly in Mercury's orbital motion and the rotating disk problem was unresolved. In time, though, Einstein convinced himself that the theory was acceptable.

OTHER GRAVITATIONAL THEORIES

The Einstein–Grossman paper and Einstein's earlier paper at Prague inspired others to work on gravity. Within a short time, three other theories were published. The first was formulated by Max Abraham of Göttingen, the second by the Finnish physicist Gunnar Nordström, and the third by Gustav Mie. A lively debate began between Abraham and Einstein, first in the journal *Annalen der Physik*, then finally in person at a conference. Abraham became quite abusive, referring to Einstein's attempts as a setback for physics. His arguments were, at times, vicious and abusive; at one point he said that Einstein's theory of relativity was "threatening the healthy development of physics." Einstein took it all in stride, but strongly defended his theory. Referring to Abraham's theory he said, "It is logically correct, but a monstrous embarrassment otherwise." On another occasion he called it "a stately horse that lacks three legs."

Einstein was, however, attracted to Nordström's theory. He said it was a consistent and logical theory, but felt that there were problems. Einstein said little about Mie's theory, mainly because it was inconsistent with his principle of equivalence and he was certain that any viable theory had to contain this principle.

TO BERLIN

Not only did Einstein's participation in the Solvay Congress introduce him to many of the leading scientists of Europe, but it also later changed his life. In early 1913, two of the participants, Max Planck and Hermann Nernst of the University of Berlin, were given the responsibility of recruiting faculty for a new institute that was to be built in Berlin—the Kaiser Wilhelm Institute for pure research. During the Solvay Congress they had become so impressed with Einstein that they decided to visit him in Zurich and make him an offer. He would be director of the Kaiser Wilhelm Institute when it was completed; furthermore, he would be given a position at the University of Berlin, and he would not be forced to teach. He could teach whenever he wished. Finally, his wages would be considerably higher than they were at Zurich. It was an offer that Einstein could hardly refuse.

Einstein was particularly attracted by the fact that he would not have to teach and could devote most of his time to research. He did not make a decision immediately, but suggested that Nernst and Planck spend the day sightseeing in the area. When they returned, he said, he would have his answer. With his usual sense of humor he told them if he was wearing a red rose when they returned the answer was yes, and if he was wearing a white rose the answer was no.

When Nernst and Planck pulled up in their carriage Einstein came to greet them wearing a red rose. He was excited about the new position, but a little unsure of himself. "The Germans are betting on me as a prized hen; I am myself not sure I'm going to lay another egg," he said shortly after he accepted the position.

On April 6, 1914, Einstein and his family moved to Berlin. His wife Mileva, however, was reluctant to leave Zurich, and relations between her and Einstein, never good at best, became even more

Einstein and Max Planck. Einstein is receiving the Planck medal. (Courtesy AIP Niels Bohr Library, Fritz Reiche Collection.)

strained. Within a few weeks of coming to Berlin, she took their two boys and went back to Zurich. It was the end of their marriage. Einstein was not lonesome in Berlin, however; he had an uncle and a cousin in the city. Furthermore, he had met a young astronomer by the name of Erwin Freundlich who was interested in verifying his theory—checking the deflection of light from stars around the sun. Freundlich had visited Einstein in Zurich and talked to him about the verification. On the evening that Freundlich arrived, Einstein invited him along to a talk he was giving and, to Freundlich's embarrassment, Einstein introduced him as the man who was going to test the new theory.

The first opportunity to check on the deflection of light came a few months after Einstein's arrival in Berlin—in August of 1914. An eclipse was to take place in Siberia. Einstein was eager to see the results, sure that his theory would be verified. Freundlich and his crew set out in late summer and, on August 1, Germany declared war on Russia; before they could take any measurements, he and his crew were taken prisoner. Fortunately, a few weeks later he was traded back to Germany for some Russian officers. Einstein was disappointed, but as it turned out, the disaster was a stroke of luck. For, within a short time, Einstein completed his theory and, using it to recalculate the amount of deflection expected, he found a considerable change.

THE FINAL STRUGGLE

Einstein was soon back at work on the theory that he and Grossman had devised. Although he had convinced himself earlier that it was satisfactory, he was now becoming increasingly dissatisfied with it, mostly because it was not covariant. He decided to bring covariance back into the theory. Once he did this everything began to fall into place. Still, a superhuman effort was required to complete it. Einstein spent two weeks locked up in his study working on the final form of the equations. During this time he spoke to almost no one, although he did write several letters to the mathematician David Hilbert of Göttingen. Einstein had delivered some lectures at Göttingen a few weeks earlier and was pleased when Hilbert took an interest in his theory.

Einstein looked again at one of the sets of equations that he and Grossman had tried earlier. Previously the two men had

Einstein enjoying his favorite hobby, sailing. (Courtesy Lotte Jacobi.)

shown that these equations did not lead to Newton's equations in the case of a weak gravitational field. Einstein soon found, though, that they had made a mistake; they did, indeed, lead to Newton's equations. He then used the new equations to calculate the orbit of Mercury and was pleased to find the result was slightly different from Newton's theory. Looking at the difference, he saw that it corresponded to a precession of Mercury's orbit. The orbit was an ellipse and, according to Einstein's theory, the major axis of the ellipse should move very slowly (precess). The amount of precession corresponded almost exactly with the known anomaly. "I was beside myself with ecstasy when I got that result," he said.

Einstein reported his results to the Prussian Academy in late 1915 and, in early 1916, he published his classic paper "The Foundations of the General Theory of Relativity" in *Annalen der Physik*.

"The laws of physics must be of such a nature that they apply to systems of reference in any kind of motion. Along this road we arrive at an extension of the postulate of relativity," he wrote near the beginning of the paper. He went on to say, ". . . we are able to 'produce' a gravitational field merely by changing the system of coordinates," implying that accelerated motion, like uniform motion, was not absolute.

Much of the first part of the paper dealt with the new tensor theory on which his theory was based. Most physicists would not be familiar with tensors and Einstein felt that he had to deal with them before getting to his equations. He then went on to derive the field equations, show how they reduced to Newton's equations in the case of a weak gravitational field, how they predicted the bending of light around a gravitating object, and the Einstein redshift. And, finally, he calculated the orbit of Mercury, showing that the anomaly was accounted for by his equations.

It is interesting that at about this same time, David Hilbert of Göttingen arrived at the same equations—with a slight difference. He was, however, helped considerably by his correspondence with Einstein.

CURVED SPACE-TIME AND MATTER

Taking a closer look at Einstein's theory we see that it differs from Newton's theory in a fundamental way. One of the things that Einstein disliked about Newton's theory was his concept of an action-at-a-distance force. According to this idea, objects such as the Earth and the sun attracted one another across empty space via a mysterious long-distance force. If you dropped a stone on Earth, for example, the effect would be felt by the moon immediately. But special relativity forbids velocities greater than that of light.

With the introduction of non-Euclidean geometry, Einstein found that space-time became curved by the matter within it. Gravity was, in essence, a curving of space-time. The matter of our sun, for example, curved the space-time around it, and Einstein

Einstein on a visit to the Mount Wilson Observatory. (Courtesy California Institute of Technology Archives.)

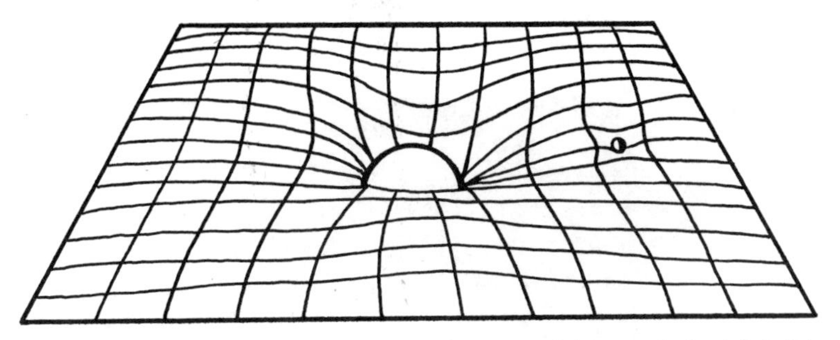

The space-time around the sun is curved by the matter of the sun. A planet feels this curvature and travels in a geodesic around the sun.

showed that anything that moved through this region (e.g., planets) would move in what is known as a geodesic (the shortest distance between two points). We know that the shortest distance between any two points in the space around us is a straight line. But if we consider the shortest distance that we can travel between, say, Tokyo and London, it is actually an arc of a circle. In the same way, the shortest distance between two points in the curved space-time around the sun is an arc of a circle, or more exactly, an ellipse. Planets, therefore, trace out ellipses as they move around the sun. But what Einstein's theory showed that Newton's didn't is that the major axis of the ellipse precesses, or changes direction, slowly (see diagram).

In the preceding chapter we talked about various properties of curved space and how we might be able to use them for space travel. We now see that space-time can, indeed, become curved by the matter within it. We cannot see this curvature directly, but we can see its effects. The extent of the curvature depends on the amount of matter present or, more explicitly, on the density of matter. An extremely dense star, for example, will curve the space-time around it much more than an ordinary star like our sun. And the closer you get to the star, the greater the curvature.

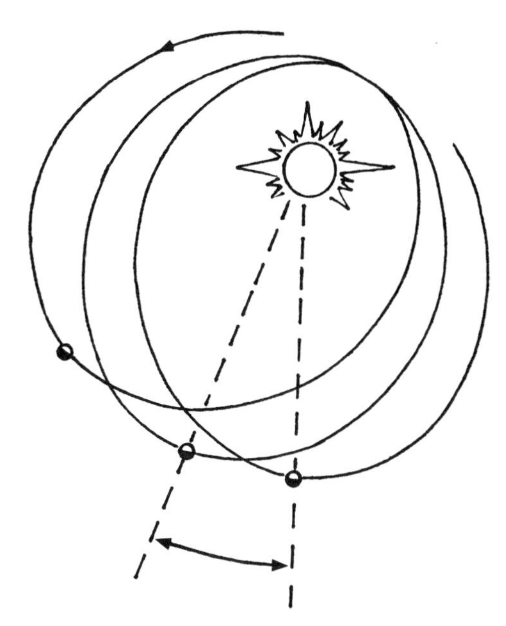

Precession of the orbit of Mercury. The direction of the major axis of its elliptical orbit slowly changes.

One of the interesting aspects of this curvature relates to the Einstein redshift. As we saw earlier this can be interpreted as a change in the rate of clocks in an increased gravitational field. This tells us that clocks will run slower and slower as the curvature of space-time increases. Is it possible that the curvature can ever be great enough so that clocks actually stop? We will explore this possibility later.

THE TWIN PARADOX RESOLVED

We are now in a position to take another look at the paradox we talked about in relation to special relativity. As we noted, time slows down relative to a fixed observer when an astronaut travels off to space, so if one of two twins travels to a distant star and back, he will remain young while his brother ages. And the greater the speed the astronaut travels, the greater the difference. The paradox was related to the fact that all motion is relative and it is therefore impossible to tell which of the twins would actually be younger when they got together again. Considerable controversy had surrounded the problem since the publication of the special theory of relativity. Once Einstein had published his general theory of relativity, however, he reexamined the question. There was, indeed, a distinct difference between the situation of the two twins. One of them would have had to accelerate away from the other and therefore would feel a force on his body. It is this twin, Einstein showed, that would actually be the younger of the two when they got back together.

In conclusion, we now know that there are two ways to slow the rate of time. It can be slowed by traveling at high speeds relative to a fixed observer and it can also be slowed by curved space. Furthermore, we know that curved space does, indeed, exist.

CHAPTER 5

The Nature of Time and Space-time

Before we continue, we should pause briefly to contemplate what we know about time. After all, time, like space, is also critical in our flight to the stars, so it's essential that we understand it as fully as possible.

Let's begin by asking the question: What exactly is time? If you presented this question to a dozen different people, you would no doubt get a dozen different answers. "It's what I read on my watch." "It's the fourth dimension of space-time." "I don't really know." "I'm in a rush and don't have time to answer," are a few of the answers you are likely to get. Scientists, in particular, will probably admit that they don't really know what it is. To them, it's something they factor into their equations, but find hard to define.

There are many ways to think of time. For years people looked upon it as a river—the "river of time." It was as if we were on a raft called "now" traveling on this river, a river that presumably flowed throughout the universe from the infinite past to the infinite future. But, if time is like a river, we are inclined to ask: Does it always flow at the same rate? You know that when you're bored, time seems to drag, and when you're doing something exciting, it goes all too fast. Is this an indication that its flow rate changes? We know, of course, that our feeling of the passage of time has little to do with the real nature of time; it's only a psychological perception.

In the case of rivers on Earth, we know that we can dam them up, in effect, stopping them. Is it possible to stop the "river of time?" Everyone, I'm sure, would like to do that once in a while; after all, it would stop the aging process, and most people don't want to grow older any faster than they have to. If we wanted to stop time, though, we would probably first have to slow it down. And, we know from an earlier chapter, that the relative time of one observer can be slowed if his speed is high relative to that of another observer. We will look into this in detail a little later.

What about reversing time? If it could be reversed, we would be able to go back to the past; we would grow younger instead of older. We will see that, although this might be possible in theory, it seems unlikely that it will ever happen in practice.

Another important issue is the question of the beginning of time. Did it actually have a beginning or has it always existed? This may seem like a strange question, but it turns out that our universe does have a finite age. It is about 18 billion years old, and, before that, as far as we know, it didn't exist. Also, what about the question of the other end of time—the destiny of the universe? Will it exist forever? And, is there a universal time, one that is the same throughout the universe?

At the present time, most of the above questions have to be answered in the negative. We certainly know of no way of slowing time down directly or of reversing it. But will it be possible in the distant future to have some control over time? This is a question that we can't easily answer right now, but it's an interesting question, one that shouldn't be ignored. Let's turn, then, to the above questions with the future in mind.

As strange as it might seem, time as we know it is a relatively modern concept. Early man was not concerned with the passage of time; he saw the sun rise in the morning and set in the evening, but short durations of time meant little to him. Similarly, young children have little sense of the passage of time. The present exists for them, but the past and future are something they do not generally think about.

Although short periods of time meant little to early man, longer time spans soon became important. Crops had to be planted in the spring and therefore a calendar was needed. The Babylonians established the first calendar in about 800 BC; it consisted of 360 days, divided into 12 months, each of which contained 30 days. A little later, the Egyptians devised a similar calendar, but they added five days of celebration at the end of the year. Of course, that still left them roughly a quarter of a day short of our actual year (the time it takes for the Earth to orbit the sun). It took man several centuries to accomplish the final squeezing in of this quarter of a day to obtain the "exact" length of the year.

The division of the day into hours was introduced by the Egyptians. They divided it into two cycles of 12 hours each, using sundials to measure the hours. The first mechanical clocks didn't appear until the eleventh century. They were large devices built in China that were controlled by slowly falling weights. A bell or chime signaled each quarter of an hour. Some of these early clocks—or at least replicas of them—are still on display today in China.

Clockmaking in Europe began about 200 years later. Most of the early clocks were used in monasteries, cathedrals, and churches and some were masterpieces of art. But, as timepieces, they lacked accuracy, sometimes losing or gaining a large part of an hour during the day.

The breakthrough that gave Europe its first accurate clocks was the pendulum. Galileo is usually credited with the discovery of its amazingly accurate period (time for one complete swing). While sitting in the Cathedral of Pisa one day, he began watching a large chandelier swinging overhead. Using his pulse, he timed its swings; they appeared to take equal lengths of time regardless of the length of the arc of swing. Returning home, he set up a small pendulum and checked the result. To the accuracy with which he could measure, he found that the period of the pendulum was independent of the size of the arc of swing. He then changed the length of the pendulum itself and found that the period was different for different lengths. He was overjoyed with the discovery.

In one respect, though, the discovery was ironic. Throughout his life, Galileo struggled to understand the enigma of natural motion and his greatest frustration was always the lack of a good timepiece. And, yet, one of the most accurate timepieces was within his reach—the pendulum. Strangely, though, he never took advantage of it. This was left to the Dutch physicist, Christian Huygens; he built the first pendulum clock 70 years after Galileo died.

With the introduction of the pendulum, clocks soon became accurate and clockmaking became a major industry in Europe. Although few clocks were privately owned, they were on public display in many places and people consequently became much more aware of the passage of time. They could "see" time pass merely watching the hands of a clock.

THE "FLOW" OF TIME

Earlier, we compared time to a river. Let's take a closer look at this idea and see if it is a true representation of how time flows. A moment's reflection will show that a river isn't a good analogy. We find, first of all, that the idea forces us back to Newton's idea of absolute time. If time were like a river, it would have to be universal and would therefore flow at the same rate everywhere. We know that this is inconsistent with Einstein's special theory of relativity. Einstein showed that time is a local phenomenon; it runs differently, or at a different rate, for different observers, depending on their motion. Two astronauts moving past the Earth at different speeds see time on Earth pass at different rates, and this is, of course, inconsistent with the idea of a universal river of time.

A similar problem exists near a strong gravitational field, or equivalently, in a region of curved space. An observer watching the clock of another observer in a strong gravitational field sees it click off seconds much slower than his own clock; in other words, he sees the other observer's time slowed down. Again, this is in conflict with the idea of a river of time.

Also, if time is like a river, we have to ask how fast this river flows. And how do we measure its flow rate? To do this, we would need a standard of comparison, and we don't have one.

Time, obviously, is not like a river, and shouldn't be compared to one. Furthermore, the idea that we are like a raft on this river makes little sense. The formulas of physics do not have a "now" in them; they deal only with time intervals. In short, time is not a physical thing like a fluid that we can examine. It's just a dimension like the dimensions of space.

CAN TIME BE REVERSED?

What about reversing time? If we want to reverse it, we must first stop it. So our first question has to be: Can we stop time? We saw that, according to special relativity, time can be slowed down. The clock of an observer moving relative to you appears to run slow, and the closer his speed gets to that of light, the slower it runs. In theory, if he moved at the speed of light relative to you, his clock would stop. But special relativity forbids a speed this high; matter can only travel at speeds less than that of light. So, from this point of view, we obviously can't stop time.

But time is also slowed down by curved space. Let's see if we can use curved space to stop it. Suppose again that we have two observers; call them A and B. Assume that they are initially in a region of flat space, then B moves into a region of curved space around a dense star. As B gets closer to the star, A will see B's clock run slower and slower. Does he ever see it stop? It turns out that he doesn't; if the curvature is sufficient, it will almost stop, but, as in the above case, it never completely stops.

It seems, therefore, that in the case of two observers comparing time rates, time can be slowed down, but it can't be stopped. I should reiterate, though, that an individual observer would never notice this slowing of time; he would be aware of it only if he compared his time to that of another observer.

This brings us to the question of time reversal. Let's see if we can use the slowing of time discussed above to speculate on how we might reverse time. First, consider an observer moving relative to you at a speed close to that of light. As I mentioned earlier, time would stop if his speed relative to you was equal to that of light. We know that this is impossible. But what would happen if he could move at a speed greater than that of light? According to the equations of relativity, his time would become imaginary (the square root of a negative number), but this imaginary time would be reversed as compared to your time. This means that at speeds greater than that of light we do get a reversal, but unless we find some way of reaching these speeds it will remain beyond the realm of science.

There is, however, an interesting reversal of time that you can see any time you want. All you have to do is look out into space, and you are looking back in time. The light you see from the stars is light emitted long ago that is just now reaching us. Although this is not quite what we are looking for, it is a reversal of time. In fact, if you look far enough, you can actually see almost all the way back to the beginning of time.

Another interesting reversal of time was pointed out by Richard Feynman of Caltech in 1949. We know that our world is made up of many different types of particles. Most people are familiar with electrons, protons, and neutrons, but there are also other types of particles such as mesons, hyperons, and so forth. In their search for new particles, physicists have found that nature has a curious symmetry: to every type of particle there is what is called an antiparticle. For example, corresponding to the electron, there is the antielectron, or positron, and to the proton, there is the antiproton. The major difference between particles and antiparticles is their charge. The electron, for example, is negatively charged while the positron is positively charged. But, of more importance, if a particle meets an antiparticle they annihilate one another. Matter disappears and in its place one or more energy particles—photons—appear. Matter is, in essence, converted into energy.

There are many different reactions involving particles and antiparticles, one of the simplest being the one that I just mentioned: the collision and annihilation of a particle and an antiparticle. We would normally tend to think of this as a particle and an antiparticle moving forward in time, and colliding. Feynman showed, however, that it can equally well be viewed as the collision of two particles, one moving forward in time, and the other one moving backward in time. What this means is, that as far as mathematics is concerned, antiparticles can be considered to be particles moving backward in time.

Another possible reversal of time that has created considerable interest in the last few years is one related to cosmology. In a later section I will talk about the beginning of the universe, in particular, the explosion called the big bang. According to the big bang theory, all galaxies were created in this explosion, and they are now expanding away from one another. In other words, the universe is expanding. One of the fundamental questions of modern cosmology is: Will the universe continue to expand forever, or will it eventually collapse back on itself? We aren't able to answer this question yet, but, if it does collapse back on itself, we have a problem: What will happen to time during this collapse? If time goes forward in the expansion part of the cycle, it is reasonable to speculate that it might go backward in the contraction part of the cycle. Tommy Gold of Cornell was the first to suggest this possibility. In such a universe, we would get younger instead of older, which seems contrary to common sense. Gold gets around this by assuming our mental processes would also be reversed, so everything would appear normal to us, and we would therefore age in the normal way.

THERMODYNAMICS AND THE ARROW OF TIME

I mentioned earlier that, as far as the equations of physics are concerned, the "flow" of time is irrelevant and makes no sense. Interestingly, the direction of time is also of no consequence. In other words, it makes no difference in our equations if we substitute negative times throughout them rather than positive times.

As far as these equations are concerned, time is symmetric, which means that the laws of science are the same whether time moves in the positive or negative direction.

Yes, but we have a psychological sense of the direction of time, you say. This is true: the past exists in our memory, and we anticipate a future, so time must have a direction. Our psychological sense of the direction of time therefore gives us an "arrow of time." But do we have anything more concrete than this? It turns out that thermodynamics, the study of heat in motion, also gives us an arrow of time.

To understand how thermodynamics accomplishes this, we must begin with some of its basic principles. Thermodynamics is based on a number of laws. Although we will mainly be concerned with the second law, for sake of completeness I will briefly mention the first law. It is nothing more than a statement of the conservation of energy, which says that energy can be converted from one form to another, but cannot be created nor destroyed.

The second law is not as easy to pin down, mainly because it can be stated in so many different, but equivalent, ways. One of the simplest is Lord Kelvin's statement of 1851: Any process that converts energy from one form to another will always dissipate some of the energy as heat. In other words, if you place a hot object next to a cool one, heat will always flow from the hot one to the cool one.

Another statement came from the German physicist Rudolf Clausius. It is based on a rather abstract concept he introduced in 1865 called entropy; he defined entropy to be the heat content of a system divided by the absolute temperature (the absolute scale is based on zero as the lowest possible temperature). Clausius found that when two bodies were brought together, heat flowed from the hot to the cool one, but he noticed that something else, something strange, was going on. The entropy of the system of two bodies was increasing. He used this to formulate the second law as: The entropy of a system which is isolated from its surroundings always increases.

Soon after the laws of thermodynamics were set up, physicists realized that there was a problem. Thermodynamics dealt with macroscopic phenomena, things such as temperature, pressure, and volume, that could easily be measured. But the basic ideas of thermodynamics applied to gases and solid matter, which consisted of atoms and molecules. It was important that the macroscopic world of pressure and temperatures be definable in terms of atomic, or microscopic, quantities.

Several scientists attacked this problem in the 1870s: the Scottish mathematical physicist James Clerk Maxwell, the irascible German physicist Ludwig Boltzmann, and the little-known American chemist Willard Gibbs. In one respect the problem was ahead of its time, since scientists were still arguing over the reality of atoms and molecules. Several well-known scientists (e.g., Ernst Mach and Wilhelm Ostwald) stubbornly insisted that there were no such things as atoms. Despite their criticism, however, the new idea triumphed and a relation between the macro and micro worlds was established. We now refer to it as statistical mechanics.

Once the ideas of statistical mechanics were established, Boltzmann took the bull by the horns and applied it to entropy, showing that it could be defined as an increase in the degree of disorder of a system. In short, entropy was a measure of the disorganization of a physical system. This meant that the second law could be stated as: the disorganization of any closed physical system always increases.

Let's consider an illustration of this version of the law. Begin by filling the bottom half of a cylindrical plastic container with red marbles and the top with blue marbles, leaving a little room at the opening. Now, shake them. What happens? Obviously the red marbles start to mix with the blue ones and the system becomes disorganized. Furthermore, it is easy to see that the more you shake the container the more disorganized it becomes. You could, in fact, continue shaking it for years and the marbles would, in all probability, never come back to their original configuration. This means that the entropy of this system never decreases and that it satisfies the second law of thermodynamics.

It is necessary, of course, that the system be closed, or

isolated from its surroundings. If we are allowed to introduce more marbles into the container, it is possible to decrease the entropy. But if we keep the number of marbles constant it will always increase. That's not to say that entropy can never decrease; it can, but "on the average" it has to increase. There are systems where there is a decrease of entropy. Your refrigerator is a good example; but it is not a closed system.

What is important about the second law of thermodynamics is that it gives us an arrow of time. It distinguishes the past from the future. An increase in entropy points in the direction of the future, or putting it another way: the future is the direction in which entropy is increasing. The past is where entropy was less.

It is important to remember, though, that entropy is a statistical concept. It doesn't exist for individual atoms or molecules. But as long as the system is closed there is no limit to what it can be applied to. Since the universe, as far as we know, is a closed system, its entropy is also increasing.

Boltzmann fought hard for his new insights. But skeptics like Mach, Ostwald, and others never let up. Frustrated, dejected, and emotionally depressed, Boltzmann finally gave up and committed suicide in 1906. Ironically, within a few years the tide began to turn and his ideas were accepted. As a memorial to his struggle his formula relating entropy and disorder was carved on his tombstone.

OTHER ARROWS OF TIME

The thermodynamic arrow of time is, perhaps, the most important one, but there are others. A second arrow, called the cosmological arrow, comes from the expansion of the universe. As I mentioned earlier, the big bang theory tells us that the universe began as a gigantic explosion. According to our best estimate, the explosion took place about 18 billion years ago, perhaps a little less. Initially, the universe consisted of a cloud of particles, but in time this cloud broke up and the individual clouds began to collapse in on themselves. They condensed finally to form galaxies, each containing billions of stars.

As we look out into the universe today, we see galaxies all around us, billions of them, all expanding away from us. The universe is expanding. This means that in the future the galaxies will be more dispersed, and in the past they were closer together. The expansion of the universe therefore also gives us an arrow of time.

The first question that is likely to come to mind now is: What is the connection between the thermodynamic and cosmological arrows? Or is there any? There is, unfortunately, a difficulty in trying to connect them and it centers on entropy. The thermodynamic arrow depends on entropy, while the cosmological arrow is based on gravity. To connect the two we have to be able to calculate the entropy of gravity—and we can't. We're not even sure the concept has any meaning.

Furthermore, there's another problem related to the cosmological arrow. It concerns the beginning of the universe. If the entropy of the universe is continually increasing, it must have been low at the beginning. But the big bang, the chaotic explosion that created the universe, seems to imply disorder. We therefore have to ask ourselves: Was the initial state one of high disorder or high order? This is still unsettled, but if it was of high disorder, we wonder if there is enough order left to allow an increase of entropy for billions of years. A theory that has generated considerable interest in the last few years may give us the answer. Called inflation theory by its creator, Alan Guth, it states that during the first tiny fraction of a second, a sudden increase in the expansion rate of the universe occurred. A number of scientists have shown that this inflation may have given considerable structure to the early universe. If so, the early universe would have started off in a smooth, ordered state, and our problem is solved.

Now, let's go to the other end of the expansion. What will eventually happen to the universe? Its final fate, it turns out, depends on how strong its mutual gravitational field is, and this field, in turn, depends on how much matter the universe contains.

If the gravitational pull is sufficient, it will collapse back on itself, otherwise it will expand forever. The consensus at the present time is that there isn't quite enough matter in the universe to stop its expansion. If so, it will continue to expand forever. But strange things happen in science and it is certainly possible that the additional needed mass may one day be found.

What will happen to time if the universe does collapse? I mentioned earlier that according to Gold, time would reverse, but our mental processes would also reverse, so we would never notice it. But what does entropy say about the reversal of time? If entropy increases as the universe expands, it is logical that it will decrease if it stops and begins to contract. In this case, the universe will eventually come back to an ordered state. Stephen Hawking and a number of colleagues, however, are not convinced that this is the case. They believe that entropy will continue to increase during the contraction. If so, the cosmological arrow will remain in the same direction as the thermodynamic arrow and time will not reverse.

This brings us to the third arrow of time. It is the one that you are, no doubt, the most familiar with. I mentioned it briefly earlier: the psychological arrow or the arrow in our mind. We have a psychological sense of the direction of time: we remember the past, we feel the present, and we anticipate the future. The direction of this arrow is, of course, the same as that of the thermodynamic and cosmological arrows. Hawking has, in fact, shown that there is a link between the psychological and thermodynamic arrows.

These are the three main arrows of time, but there are some others that are worth mentioning. First, there is the electromagnetic arrow. Electromagnetic waves, of which ordinary light is one form, propagate forward in time. Stars emit radiation and this radiation moves out from their surface. Light from our earth is also moving out into space, as it is from all objects. Furthermore, it has a finite speed. This is why when we look into space, we are looking back in time.

When we send a message to a spaceship out in space, it travels into the future and therefore has a particular direction. If this were not the case, we would be able to send signals into the past and this would mean that we would be receiving signals from the future. And we know, of course, that we aren't.

Finally, there is a microscopic arrow associated with nuclear physics. Somehow it seems out of place and we're not sure why it exists. It is well known that virtually all nuclear reactions are time symmetric. For example, if we have a reaction $A + B \rightarrow C + D$, we also have the reaction $C + D \rightarrow A + B$. In some cases, the reverse reaction is rare; neverthless it occurs. In one case, however, it doesn't and that is the case of the decay of the *K* meson. Most of the time the *K* meson decays to three particles in a time-symmetric reaction, but occasionally it decays to two. This reaction is not time symmetric and no one understands why, but because of this, it can be used as an arrow of time.

UNIVERSAL TIME

Earlier I talked about the age of the universe; I said it was about 18 billion years old. But does this really make any sense? Eighteen billion years by whose watch? After all, we know that time is local; the rate at which it runs depends on the relative motions of the observers. This means that two observers moving at high speed relative to one another see the universe age differently.

Does this mean that the universe only has a relative age, or an "individual" age, for each observer? Is there no "real" or absolute age to the universe? It turns out that we *can* talk about an absolute age. It can be defined as the time that an observer would experience if he were moving with the expansion of the universe. And since the velocity of recession of the galaxies is only a small fraction of the speed of light, this "universal" time corresponds closely to Earth time. In fact, it is the time experienced by any of the galaxies. Therefore, it makes sense to talk about a universal time. That's not to say, though, that everyone experiences this time. If you are traveling at a high speed relative to the expansion or if you are in a strong gravitational field, you will experience a time dilation and the age of the universe will be different from the one we experience here on Earth.

TIME AT THE ENDS OF TIME

At the beginning of this chapter we posed the question: Was there a beginning to time? Most people would answer this by saying that time began when the universe began, in other words, 18 billion years ago. This is true—as far as our present universe is concerned. But what if our universe eventually collapses? This means that it might reexpand, which in turn implies that our universe might be an oscillating one with many cycles. Time would certainly have existed during these earlier cycles.

Furthermore, if we look carefully at the concept of time during the very early universe, we find another problem. To understand it, it is best to think of the expansion of the universe in reverse, in other words, in a state of collapse. The galaxies will all come together and merge until, finally, the universe is a sea of gas—radiation and particles. As this gaseous cloud is squeezed down, its density gets higher and higher. Finally, it will be so high, and its volume so small, that general relativity is no longer applicable to it. It has, in effect, broken down. This point occurs at 10^{-43} seconds after the big bang (or 10^{-43} seconds before the big crunch in a collapse); it is called the Planck time.

The reason why general relativity breaks down is that we are now in the realm of atoms and molecules, the quantum realm, and a quantized version of general relativity is needed to explain things. At the present time, we don't have one. We can therefore say nothing about what would happen to time. John Wheeler of Princeton and others, however, have speculated on what this region would be like. They visualize it as a sort of foam like soap bubbles. Space-time is broken up into regions of space, time, and nothing. Time is therefore completely distorted on this scale and may have no meaning.

Of course, if we talk about what happens to time at the beginning of time, we should also talk about what happens to it at the end of time. Unfortunately, there's not much to say. If the universe continues to expand forever, time will just go on forever.

SPACE AND TIME

With this, we come back to the question of how we should view time and what it really is. We have seen that it doesn't flow in the usual sense and it can't be reversed. Furthermore, it is local and depends on the location and relative speed of the observer. Nevertheless, it is possible to define a universal time. Finally, we have seen that there are problems at the beginning of time and in considering extremely small intervals of time.

It is important also that we treat time as a component of space-time. Space and time are not independent. The curvature we talked about earlier is one of space-time, not just space. We can, in theory, separate space and time and talk about them separately, but no absolute significance can be attributed to them separately.

How, then, should we define time? General relativity, which treats time merely as a dimension, seems to provide the best, and the most complete, description. It does not, however, answer all of our questions. Many problems remain, and it may be many years before they are solved. Our survey of time here has, nevertheless, given us a better understanding of its idiosyncrasies, and we are therefore now in a better position to look at how time and space-time are affected when they undergo extreme warping.

CHAPTER 6

The Discovery of Space-time Tunnels

We have seen that matter curves the space around it and that the greater the amount of matter, the greater the curvature. But how extreme can the curvature become? Is it possible for matter to create an infinite curvature? This possibility bothered Einstein and others after the first solution of the field equations of general relativity was found. Indeed, a mystery developed that was not cleared up for over 40 years. It was, however, the final resolution of this mystery that now forms the basis of our hope for space travel.

THE ACCEPTANCE OF GENERAL RELATIVITY

The mathematical structure and physical interpretation of Einstein's theory differed significantly from Newton's theory of gravity. For one thing it was much more mathematically complex. Still, the predictions of the theory could not be too different from those of Newton, for it was well known that Newton's theory gave excellent results for moderate gravitational fields. The motion of the planets in the solar system, for example, was predicted to a high degree of accuracy.

Einstein's equations, however, predicted a slightly different orbit for Mercury. They told us that the major axis of Mercury's orbit changed direction slowly and this had been verified. But a verification of the bending of light around the sun still eluded Einstein. And with World War I now in progress, it seemed that there was little chance of checking on it for some time.

At this stage, Einstein's theory was still generally unknown outside of Germany and with the war on there was little communication between Germany and the outside world. A copy of it did, however, reach Willem de Sitter in Holland and he passed it on to Arthur Eddington in England. After studying it carefully Eddington became intrigued with the theory. Over the next few years, he wrote about it extensively; furthermore, he brought it to the attention of the Astronomer Royal of England, Frank Dyson, pointing out that a verification of the deflection of light could be made if stars were visible near the sun during a total eclipse.

Dyson noticed that an eclipse on May 29, 1919, which was expected to occur in the constellation Hyades, would be surrounded by several bright stars. It would afford an excellent opportunity to check on the validity of the theory. But at this stage (1917), England was still at war with Germany. To most it seemed incomprehensible that English scientists would be interested in verifying a theory that had been published in Germany. Eddington, however, was a Quaker and a strong pacifist, and he felt that there should be no barriers to scientific research, wars included. As it turned out, the eclipse expedition, and subsequent newspaper reports, did help heal some of the wounds of war.

Eddington and Dyson began preparations for the expedition. The eclipse path would be mainly over the Atlantic Ocean, but parts of it crossed the west coast of Africa and South America. Two sites were selected, one in each of the two countries. The first was on Principe Island in the Gulf of Guinea and the second was in Sobril, Brazil.

In January of 1919, several photographs were made of the region of Hyades where the eclipse would take place. The stars in these photographs would be compared to those seen around the sun during the eclipse. Einstein had predicted that they would be deflected from their normal position by 1.74 seconds of arc.

In April, final preparations were made and Eddington and his

Sir Arthur Eddington.

crew departed for Principe. Two other astronomers and their group departed for Sobril. Eddington arrived in Principe a full month before the eclipse, but a lot of hard work followed. A camp had to be set up, instruments tested, test photographs had to be made, and so on.

The day of the eclipse finally arrived and Eddington was shocked when he woke up. Heavy rain was pounding the outside of the tent. Despite the weather, Eddington and his crew went ahead with preparations. Finally, at about noon the rain stopped and the clouds began to break up. The sun was not visible until slightly after 1:00, and by then the eclipse had already begun. With the first view of the sun, however, Eddington swung into action.

The eclipse itself was spectacular, with a huge brilliant prominence, a giant flamelike projection from the surface of the sun, visible at totality. But Eddington hardly saw it; he was too busy taking plates. He took 16 in all, but he knew that the first ones would probably be useless; large numbers of clouds still lingered around the sun at this stage. He pinned his main hopes on the last six.

Without stars in some of the photographs, the expedition would be a total loss as far as Eddington was concerned. He had his heart set on verifying Einstein's theory and would have been tremendously disappointed if he didn't succeed. When the eclipse was over, Eddington was so anxious to see the results that he began developing the plates the very next evening. For six nights, Eddington and his team developed plates, then he began measuring them. The first ten were useless; no stars were visible because of the clouds. On the last six, stars could be distinguished, but only one of the six was really good, with several bright stars visible. Eddington measured them. He knew that field measurements such as these were crude and would have to be redone back in England, but at least they would give him an indication of whether or not there was a deflection. When he had completed his measurements, he was relieved; there was a deflection and it was relatively close to Einstein's prediction. It appeared to be slightly less, but it was certainly within experimental error. He was sure that more accurate measurements would yield better results. "It was the greatest moment of my life," he said later.

But Eddington had only gotten one really good plate. He was anxious to find out what the Sobril group had obtained. Upon his return to England he helped develop the Sobril plates. Again, the first few plates proved disappointing, but on the last half dozen, stars were clearly visible. And when the measurements were completed the verdict was the same as before: Einstein's prediction was verified (within experimental error).

Eddington made a preliminary announcement of the discovery at a Royal Astronomical Society meeting in September, but was still cautious at this point, saying that the plates would have to be checked more carefully. A colleague of H. A. Lorentz of Leiden was in the audience and when he returned to Holland he told Lorentz of the announcement.

At this stage, Einstein knew that an expedition had been sent to check on his prediction, but he knew little else. Even though the war was over, communication between England and Germany was still almost nonexistent. In September, Einstein began to get a little impatient; he wrote a friend telling him he had heard nothing and that he was getting worried. Then, a few weeks later, he wrote to Felix Ehrenfest in Holland, asking him if he'd heard anything. Ehrenfest passed the letter on to Lorentz, who had just heard from his colleague. He immediately telegraphed Einstein, "Eddington found star displacements at rim of sun. . . ."

Einstein was overjoyed. He wrote to his mother, "Good news today. H. A. Lorentz has wired me that the British expeditions have actually proved the light shift near the sun." Einstein then traveled to Holland to see Ehrenfest and find out if he could get more details. Just before he left, he got another telegram saying that Eddington's preliminary measurements had verified the prediction approximately and that he wanted to do more accurate measurements.

It's interesting that even though Einstein was anxious about the results his confidence in his theory never wavered. He was discussing his theory with one of his students when she asked him about the verification. Einstein handed her Lorentz's telegram. "This will perhaps interest you," he said. After she had read it, she asked what he would have done if it had not been verified. "Then I would have been sorry for the dear Lord—I know the theory is correct," he answered.

While Einstein was in Holland, the Dutch Royal Academy met in Amsterdam and Einstein was invited. He was warmly welcomed by the president and an announcement was made that preliminary test results from the eclipse expedition had verified his theory. News of the result quickly spread throughout the scientific community in Europe. But the real impact of the discovery didn't come for another ten days, when Eddington finally completed all of his measurements.

On November 6, 1919, the Royal Astronomical Society met at Burlington House in England. The announcement of the verification was made by the Astronomer Royal, Frank Dyson. Dyson concluded his brief statement with, "After careful study of the plates I am prepared to say that they confirm Einstein's prediction. A very definite result has been obtained, that light is deflected in accordance with Einstein's law of gravity."

According to Whitehead, who was in the audience, "The whole atmosphere of tense interest was exactly that of a Greek drama." Eddington then rose and added a few words to Dyson's announcement. But the most dramatic statement of the evening came from Nobel Laureate J. J. Thomson, the chairman. "This is the most important result obtained in connection with the theory of gravity since Newton's day . . . it is one of the highest achievements of human thought," he said.

Once the announcement had been made, news spread rapidly, and on November 7, 1919, Einstein awoke to find himself a worldwide celebrity. Headlines in papers read, "A New Revolution in Science." According to the reports, Newton's theory had been overthrown and replaced by an astounding new theory, general relativity.

Hundreds of reporters arrived at Einstein's house. He was overwhelmed and didn't know what to do; he hadn't expected such a reception. But whether he wanted it or not, it was there; he was now the most famous scientist in the world. Only once before had the press, and the world, gotten so excited about a scientific discovery, and that was in 1895 when Roentgen discovered X rays. In that case, most of the attention was directed at the discovery. Strangely, in this case, most of the attention was directed to the man. Who was he? The theory was so complex that little was said about it, except that almost no one could understand it.

Einstein didn't like all of the publicity; it severely cut into his

Einstein in later life. At the time he was working on his unified theory. (Courtesy Lotte Jacobi.)

research time. But there was little he could do about it. He soon had requests from countries around the world to speak and over the next few years he did a considerable amount of traveling. Although he continued to publish, his most productive years were now over. He tried for another 30 years to extend his theory to electromagnetic theory, to create a unified theory of gravity and electromagnetism, but he failed.

THE SCHWARZSCHILD SOLUTION

When Einstein published his theory of general relativity in late 1915, he only had an approximate solution to the equations. But his paper soon came to the attention of an astronomer stationed on the Russian front, Karl Schwarzschild. Despite the cold

Karl Schwarzschild.

and his failing health, Schwarzschild was able to find an exact solution within a short period of time.

Born in Frankfurt in 1874, Schwarzschild became interested in astronomy at an early age, writing a paper on the orbits of double stars at the age of 16. Soon after he received his Ph.D. from the University of Munich, he accepted a position at Göttingen. At Göttingen he made several important contributions to astronomy: he developed a method for measuring the light intensity of variable stars, he contributed to the theory of stellar evolution, and finally he worked on spectroscopy, quantum theory, and general relativity. He was different from most astronomers in that he was an excellent mathematician.

He had been interested in the shape of the universe for many years. Shortly after he got his Ph.D., he published a paper on the topic titled, "On the Admissible Curvature of the Universe." In it he discussed possible geometries of the universe and suggested several possible types of curvature. Later, in 1914, he tried to observe the gravitational redshift Einstein had predicted in 1911, but was unsuccessful.

As a scientist, he could have avoided serving in the army, but when World War I started, he enlisted and was sent to an artillary unit to study the trajectories of long-range shells. While serving there he developed a rare and serious skin disease called pemphigus. Despite his condition, the extreme weather, and the war around him, he worked on Einstein's theory and arrived at a solution in December 1915. His first paper described the "exterior" solution of a point of mass in empty space; in other words, it gave the curvature of space around the point. This is still one of the most important solutions of Einstein's theory, and for many years after its discovery it was the only known solution. Schwarzschild sent his result to Einstein in Berlin.

On January 9, 1916, Einstein replied, "I have read your paper with the utmost interest. I had not expected that one could formulate the exact solution of the problem in such a simple way." A few days later, Einstein communicated the solution to the Prussian Academy.

But Schwarzschild wasn't finished. His first paper gave the solution for the space around a point particle. He now considered a spherical mass and determined the curvature inside the matter; this is referred to as the "interior" solution. This solution would apply, for example, to the inside of a star.

In February of 1916, he sent this second solution to Einstein. He tried to continue working on the theory, but his condition worsened, and two months later, on May 11, 1916, he died.

Schwarzschild's solution was strange and, in many ways, troubling. The difficulty centered around what is called its "singular" behavior. A singularity is a region where the theory breaks down, in other words, the answer goes to infinity. Most theories have singularities, but they usually occur at the origin. For example, there is a singularity in the electric field around a point charge. At the origin—the position of the point charge—the electric field goes to infinity. A similar situation occurs in Newton's gravitational theory. So something of this sort was expected in Einstein's theory.

Schwarzchild showed that there was a singularity at the position of the point mass in Einstein's theory, but, strangely, he found another singularity farther out, at a small finite radius that depended on the magnitude of the point mass. He commented on it but did not try to explain it. In his second paper, the one giving the interior solution, he calculated the radius at which this singularity occurs for the case of the sun, getting a value of 3 kilometers. This radius is now referred to as the gravitational radius.

It is obvious that this singularity bothered Schwarzschild, because it meant that he could not get a proper solution inside the gravitational radius. The region inside this radius appeared to be completely cut off from the outside world. He showed, however, that a static sphere of fluid of uniform density could not be compressed inside its gravitational radius; the pressure at the center would become infinite before the gravitational radius was reached. This seemed to imply that the region inside the gravitational radius was inaccessible and could be ignored. Still, it was puzzling to Schwarzschild. Why would a "hole" exist at the center of the curved space?

A more detailed look at this "hole" was made by a 30-year-old Dutch student named Johannes Droste who was working on a Ph.D. thesis under H. A. Lorentz. Droste independently obtained the same exterior solution that Schwarszchild had, but he went on to examine the trajectories of particles and light rays in the space around the central mass. He noticed that the light rays would be severely bent in the strongly curved space, and at a radius 1.5 times the gravitational radius (the radius of the hole), the rays would take up circular orbits. Of even more interest is that he found that particles falling into the central mass would never reach the gravitational radius (as seen by an outside observer). The particles would just get closer and closer, but even after an infinite time they would not reach it. To Droste this meant that the hole at the center was inaccessible. There was no way it could be reached, so why worry about it.

Still, the situation was bothersome. Later in the year (1916), Ludwig Flamm decided to take a closer look at the geometry of the solution. If you could draw a surface representing the curvature of space around the point particle, what would it look like, he asked himself? He found that the curvature was funnel-shaped and terminated in a circle at its narrow end.

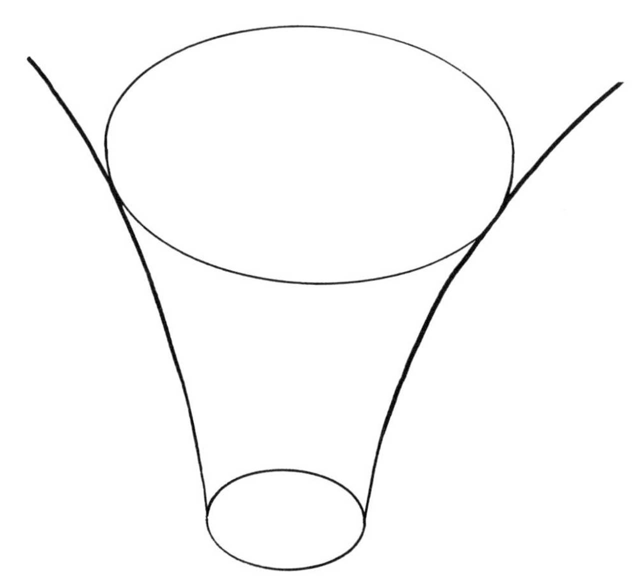

A plot of the curvature around the "invisible sphere" found by Flamm. The invisible sphere is shown at the bottom. The radius of this sphere is the gravitational radius.

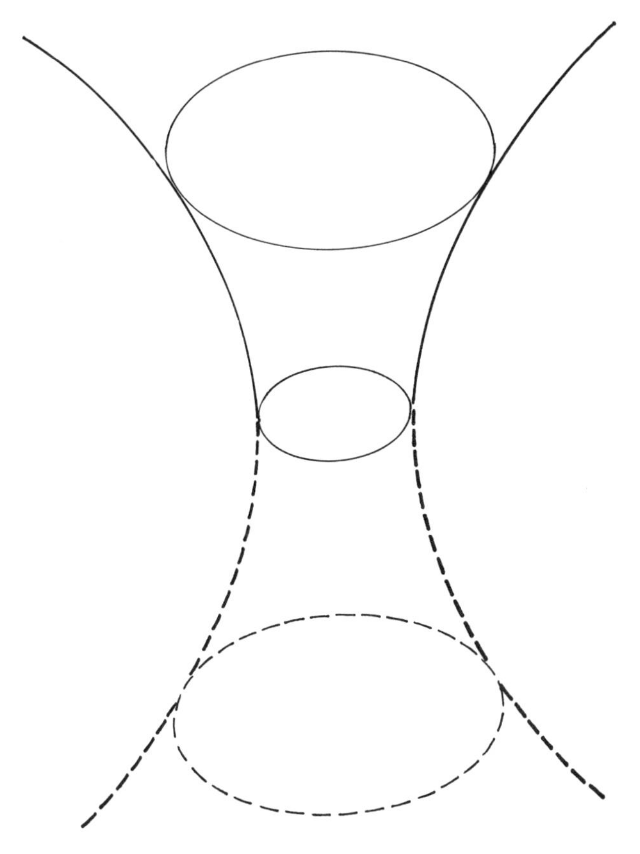

An extension of Flamm's plot found by Herman Weyl.

Herman Weyl of Zurich took this one step further and extended it beyond the central particle. It then looked like a two-sided funnel. But what did it mean? Was there any significance to this second funnel?

Most people working in the area were now convinced that the entire region within the gravitational radius should be ignored.

Several scientists, in fact, had shown that "every known mass—even the nuclei of atoms—had radii greater than their gravitational radius."

But not everyone agreed that the region should be ignored. David Hilbert of Göttingen was sure something was missing or misunderstood. His fellow countryman Cornelius Lanczos was also unconvinced; he showed that the Schwarzschild singularity could be removed by a change or transformation of the coordinates. But in using this transformation Lanczos obtained another singularity at a different position. So it didn't help much. Still, it was surprising to him that he could change the position of the singularity.

What did Einstein think of the dilemma? Before 1922 he said little. During a conference in Paris in 1922, however, the French mathematician Jacques Hadamard asked him what he thought. Einstein replied, "If that term could actually vanish somewhere in the universe [corresponding to a mass completely inside the gravitational radius], it would be a true disaster for the theory; and it would be very difficult to say 'a priori' what could happen physically because the formula does not apply any more." Einstein was obviously convinced that no mass could exist inside its gravitational radius. He later referred to the problem as the "Hadamard disaster." Hadamard's question, however, did bother him and he made a calculation to prove to himself that it was impossible for any physical system to reside completely inside its gravitational radius. He announced his result at the meeting the following day.

But this didn't end the dilemma. Eddington soon entered the picture. He referred to the region inside the Schwarzschild radius as the "magic circle." One way of getting around the problem, he said, was to envision an impregnable particle laying inside the Schwarzschild radius. And for many years discussions centered around this mysterious particle.

Others—Painleve, Jean Becquerel, Carlos De Jans, Herman Weyl, Von Laue—to mention only a few, worked on the problem after 1922, but made little progress. For ten years scientists struggled to explain the enigma. Then, in 1932, an important breakthrough was made, a breakthrough that was narrowly missed earlier by both Eddington and Weyl. Georges Lemaître of Belgium showed that the "Schwarzschild singularity" wasn't a singularity after all. Lemaître reported his discovery in a long and complex cosmology paper, however, and it attracted little attention. But it was noticed by the American cosmologist H. P. Robertson of Princeton. Realizing that the surface associated with the gravitational radius wasn't a singularity, Robertson took another look at Droste's calculation of particle orbits around it. He noticed that Droste had considered only the time of an external observer, someone watching the event from a distance. But the clock of an observer falling into the gravitational radius would run at a different rate. Robertson therefore considered the fall of a particle from the point of view of somebody moving alongside the particle and found that it would, indeed, reach the gravitational radius. The magic circle was accessible after all.

Robertson talked about his discovery in a lecture in Toronto in 1939. In the audience was J. L. Synge of Ireland. Synge told Einstein of Robertson's discovery and Einstein became alarmed. He was sure the region was inaccessible and went off to prove it. Considering a dense cluster of stars in space, something like a globular cluster, he examined the orbits of the stars around its center and found that for a distance less than 1.5 times the gravitational radius of the cluster, the stars would have to have a speed greater than that of light. Since matter couldn't travel that fast, he concluded that an object with all of its mass inside the gravitational radius couldn't exist.

There was, however, a flaw in Einstein's argument. He considered only a static distribution of matter, a distribution of stars with circular or elliptical orbits. But while he was making the calculation, Robert Oppenheimer of Caltech and a student of his, Hartland Snyder, were considering a nonstatic distribution of matter—a collapsing star. And they found that the matter could fall inside the gravitational radius.

THE EINSTEIN–ROSEN BRIDGE

Four years earlier, in 1935, Einstein and a colleague, Nathan Rosen, had tried another way to get around the problem of the Schwarzschild singularity. Their approach would have important and long-ranging implications. In this paper, they considered the problem from a geometric point of view, as Flamm had years earlier. Following the curvature down to the gravitational radius, Einstein and his colleague found that the funnel-like surface led to a circle with a radius equal to the gravitational radius. Examining the equations in detail they saw as Weyl had earlier that a similar surface was attached to the "back end" of the original surface. Both of these funnels led to a flat Euclidean space. Einstein and Rosen decided, therefore, that the best representation of this solution was two flat symmetric spaces joined by a kind of bridge, or tunnel. The point mass (also called the Schwarzschild particle) would be at the center of the narrowest part of the tunnel. One of the flat spaces represented our universe and the other represented "another universe." The two universes were, in effect, separated by a bridge, what was later called an Einstein–Rosen bridge.

Einstein was worried about the reality of this "other universe," and was relieved when he found that it would take a velocity greater than that of light to pass through the bridge to the other universe. It was therefore cut off and completely inaccessible to us and we didn't have to worry about it.

Still, there were problems with the Einstein–Rosen representation. Many were skeptical of this "other universe." John Wheeler of Princeton University became interested in it and decided it would make more sense to assume that the bridge led to another point in our universe rather than to a different universe. He began referring to the bridges as wormholes in space. He then developed a theory called geometrodynamics around the idea. Working with Charles Misner of the University of Maryland, he showed that small wormholes of this type could be used to represent particles. Furthermore, using these wormholes he found that he could

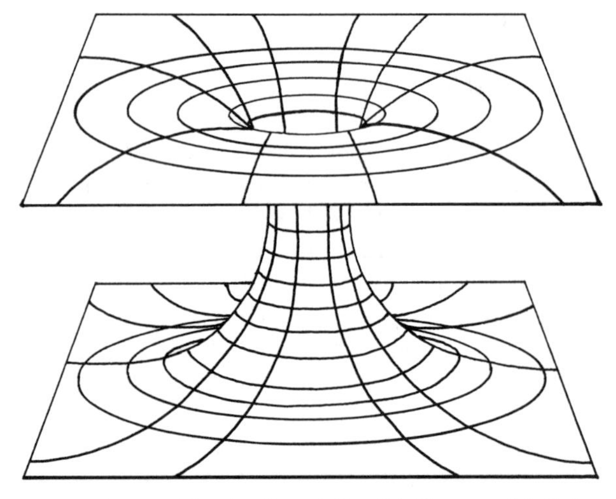

A simple representation of an Einstein–Rosen bridge. It is like a wormhole that connects two different universes.

incorporate electromagnetic theory into general relativity. In this case the wormholes would have electric field lines associated with them. An electron, for example, could be represented as a wormhole with an electric field entering and passing through it. Its antimatter particle, the positron, would lie at the other end of the wormhole where the electric field lines exited (see diagram).

Shortly after Wheeler and Misner put forward their gravitation–electromagnetism unified field theory, they found that it had actually been discovered many years earlier (1925) by G. Y. Rainach. Rainach had not used wormholes in his theory; nevertheless, the mathematics was unmistakably the same. For several years there was considerable interest in geometrodynamics, but it eventually developed difficulties, and few, if any, scientists now accept it as a true unification of gravitation and electromagnetism. The important point, though, is that space around the Schwarzschild particle was now considered to be severely curved into the form of a tunnel, or wormhole, through space; and the size of

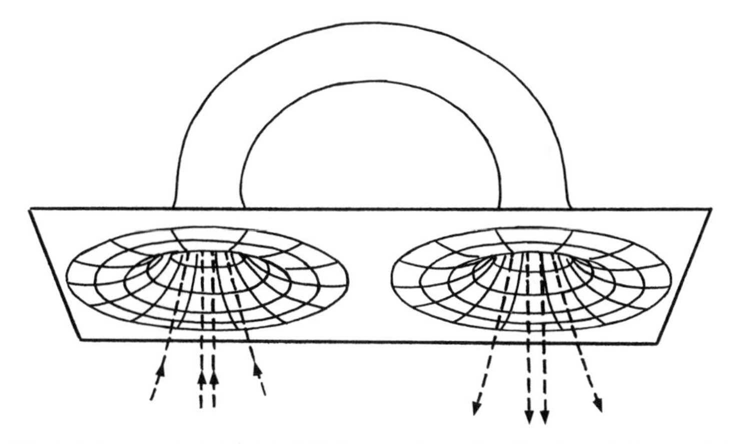

Wheeler's tiny wormhole. Electric field lines are shown threading through the wormhole.

the tunnel depended only on the mass of the Schwarzschild particle.

KRUSKAL COORDINATES

Scientists now knew that there was a tunnel through space (or, more properly, space-time) associated with the Schwarzschild particle, but they still didn't have a proper way of representing it geometrically. What was needed was a representation that showed the space properly and clearly illustrated that the "Schwarzschild singularity" wasn't a singularity. The crucial step was taken by a plasma physicist at Princeton University by the name of Martin Kruskal. Wheeler's work on general relativity had come to the attention of Kruskal and he decided to learn something about it. He therefore persuaded several other people at Princeton to form a small study group to go through one of the standard texts on the theory. Upon examining the region around the gravitational radius, Kruskal found that by using a mathematical transformation (change in coordinates) he could represent it in a unique way, a way that illustrated much more clearly what was really going on (see figure on page 139). He showed his calculations to Wheeler, but Wheeler showed little interest in them, so he decided not to publish them. But as Wheeler continued working on general relativity he soon came to realize that Kruskal's transformation was of fundamental importance. He therefore presented it at a conference in 1959, then wrote it up in a paper in 1960, giving full credit to Kruskal. As we will see in later chapters, Kruskal's coordinates are now used extensively in discussing the curved space around the gravitational radius.

DETAILS OF THE WORMHOLE

The wormhole through space discovered by Einstein, Rosen, Wheeler, and Misner was an intriguing object. It seemed to join two different spaces, but, according to Wheeler, they were two different points in our universe. Yet, there appeared to be no way we could actually travel through this wormhole. Einstein showed that, to get through, a speed greater than that of light would be needed. Furthermore, there was another obstacle. When Wheeler began examining the dynamics of the wormhole, in other words, how they changed in time, he found that they pulsed. In short, the throat of the wormhole narrowed as time passed, eventually cutting off access through the hole. Later, it would open again.

These space-time tunnels, or wormholes, were, however, a theoretical entity predicted by general relativity. Did they actually occur in nature? And if so, how? We know that matter curves space, but how would we get a wormhole? Their link to reality, we will see, is the black hole.

CHAPTER 7

Rips in the Fabric of Space: Black Holes

Black holes curve the space around them, and if you entered this space, it would seem as if you were entering a tunnel. There would be an opening at one end and an exit at the other. But is it possible to travel through these tunnels? Before we answer this question we must first examine black holes themselves, the objects that create the tunnels. What exactly are black holes? How are they formed? To understand them properly, we must begin with objects known as white dwarfs. For many years white dwarfs were the exotic objects of astronomy: tiny, bright stars, hardly bigger than Earth, yet so dense that a teaspoon of material from them weighed tons.

THE INVISIBLE PUP

The first white dwarf was discovered by the German astronomer Friedrich Bessel in the year 1844. Bessel had just determined the distance to the nearby star 61 Cygni using the phenomenon of parallax. The best way to understand parallax is to hold your finger up in front of you and alternately blink your eyes. Notice how your finger moves relative to a distant wall. That is parallax.

Bessel took photographs of the star 61 Cygni from two different positions of the Earth's orbit, 6 months apart, and noted that it appeared to move slightly relative to background stars. By measuring the associated angles he was able to determine the distance to 61 Cygni.

With the completion of this work, Bessel turned to Sirius, the brightest star in the sky. As in the case of 61 Cygni, he plotted Sirius's motion through the sky. This time, however, he found a strange "wobble" in its motion; the plot was S-shaped. To Bessel this indicated that Sirius had a companion, a tiny neighbor that was so dim it was beyond the range of his telescope. Bessel didn't worry about it, but it did seem strange to him that an invisible star, or perhaps a planet, could cause such a large wobble.

In mythology, Sirius is the dog of Orion, and hence is usually called the dog star. The invisible companion was therefore soon referred to as the invisible pup. For 20 years no one saw Bessel's invisible pup. Then, in the mid 1860s, the American lens maker Alvin Clark decided to test a reflecting telescope he had just completed by looking at Sirius. To his surprise he found a tiny white speck off to one side of it. He thought, at first, that it might be a defect in the telescope, but when he checked other stars he found that there was no indication of a problem. Sirius actually did have a small companion.

What was particularly surprising, though, was its color—white. If it was a companion to Sirius, and not a background star, it was nearby, only about 8 light-years from Earth. Dim stars at this distance are usually red. The object puzzled astronomers for years, but no one looked into it further. Then, in 1915, Walter Adams of Mount Wilson Observatory obtained its spectrum (array of lines seen when light is passed through a spectroscope), and a much greater puzzle emerged. Its spectrum indicated that it was extremely hot, just as hot as Sirius. Furthermore, with its intensity known, astronomers could calculate its size: it wasn't much larger than the Earth. This amazed scientists; a star this small had never been seen before. But it was only a preview of what was to come. The tiny white star was exerting a tremendous tug on Sirius and from the wobble astronomers could determine its mass. The result was bizarre; this star was denser than anything that had ever been discovered. A pint of material from it, if brought to Earth, would weigh 25 tons.

Most astronomers thought a mistake had to have been made. It seemed impossible. But at this time little was known about the structure of stars. Shortly thereafter, in 1916, however, Eddington began his pioneering work on stellar interiors and soon astronomers realized the problem was more serious than they had anticipated. Eddington showed that stars are gaseous throughout, and are stable because they are balanced between two strong forces: an enormous gravitational pull and an outward gas pressure. But white dwarfs didn't fit into Eddington's theory; they were different, with densities hundreds of times greater than ordinary stars. This seemed to indicate that the atoms within them were packed much closer together than in ordinary stars. Was this possible?

Eddington realized that if these objects were as dense as they appeared to be, another test could be made. Einstein's theory of relativity predicted a redshift of the spectral lines from a star and the magnitude of the redshift was proportional to the density of the star. Eddington wrote to Adams asking if he could measure this shift. Adams checked and, indeed, found a large redshift, consistent with a high density. This seemed to clinch things; Eddington was now convinced that the stars had to be exceedingly dense, but there were other problems. It was possible that a compacting of the atoms had taken place if it was assumed that the star was exceedingly hot. But when the star cooled off, Eddington was sure it would have to expand back to its original size, and this meant it had to do considerable work against gravity. Was this possible? Eddington was sure that it wasn't.

Finally, in 1926, there was a breakthrough. R. W. Fowler of Cambridge University applied quantum theory to Eddington's equations of stellar structure and showed that in addition to the usual gas pressure within stars, there was another pressure that became important when the star collapsed. It occurred because the electrons could be squeezed together only so close and no closer. He called it degenerate pressure.

Fowler's work was taken up by a 19-year-old Indian student

named Subrahmanyan Chandrasekhar. While on the long sea voyage to study at Cambridge University, Chandrasekhar decided to look into the details of Fowler's work. Within a short time he found that Fowler had overlooked something; the particles inside dense stars would be moving at exceedingly high speeds, so high that relativity theory would be needed. He applied relativity and found something strange: a massive star could collapse and become a white dwarf, but there was a limiting mass. A star would become a white dwarf only if it had a mass less than 1.4 solar masses. Chandrasekhar showed his strange result to Fowler when he got to England, but Fowler was skeptical. Chandrasekhar rechecked his calculations, but he couldn't find anything wrong with them. He therefore delved further into the theory of white dwarfs and over the next few years published several papers on them, papers that are now considered to be classics. Strangely, though, he was challenged in his efforts by Eddington. Eddington scoffed at his results, calling them nonsense. This discouraged Chandrasekhar and, because of it, he eventually discontinued working on white dwarfs.

NEUTRON STARS AND LIFE CYCLES

For years white dwarfs held center stage as the exotic objects of astronomy. It seemed impossible that such a strange object could exist, but there appeared to be no way around them. Then, in 1932, the neutron, the neutral particle of the atomic nucleus, was discovered, and within a short period of time there was speculation that an even more exotic stellar object might exist—a neutron star.

Several days after the announcement of the discovery of the neutron, Niels Bohr, Lev Landau, and Leon Rosenfeld were discussing the new particle. During their discussion, Landau suggested that some types of stars might be composed entirely of neutrons; strangely, though, he didn't published the suggestion until several years later. But how would these neutron stars form?

Fritz Zwicky of Mount Wilson Observatory was sure he knew the answer. The key was the supernova explosion (the explosion that occurs when a massive star dies). In 1933, Zwicky, along with colleague Walter Baade, published a paper making the following suggestion.

> With all reserve we advance the view that a supernova represents the transition of an ordinary star into a neutron star, consisting mainly of neutrons. Such a star may possess a small radius and an extremely high density.

Calculations soon showed that neutron stars, if they existed, would be hundreds of times denser than white dwarfs. Furthermore, they would be much smaller.

Zwicky was anxious to work out the details of the internal structure of such a star, but was beaten by Oppenheimer and his student George Volkoff. Volkoff and Oppenheimer applied Einstein's theory of general relativity to a collapsing star and found that it could end as a neutron star. Their paper, which was published in 1939, was titled "On Massive Neutron Cores." They found that neutron stars would be formed and they would be far denser and smaller than white dwarfs. There was some interest in their results initially, but most astronomers considered them highly speculative, and they were soon forgotten. Not until after the discovery of pulsars (pulsating stars of very short period) 23 years later did astronomers take another serious look at neutron stars.

Despite the detours, things were now beginning to make a little more sense. The exotic stars, white dwarfs and neutron stars, appeared to be associated with the death of large stars. Stars, it seemed, went through a life cycle: they were born, lived out their lives, then died. They were born when large gas clouds condensed, producing temperatures on the order of 15 to 20 million degrees at their center. This triggered a thermonuclear furnace that produced an outward force which counteracted the inward pull of gravity. Throughout most of its life, a star remained in equilibrium, with these forces balanced. Little happened to it; it just peacfully burned its fuel. During its early life this fuel was hydrogen, but later on it was heavier elements such as helium, oxygen, and carbon. Eventually, though, the star ran out of fuel and gravity began to overcome it. The final state of a star, whether it became a white dwarf or a neutron star, depended on its mass. There is, however, a complicating factor: stars lose considerable mass during their lifetime, and, of course, considerable mass is lost during the supernova explosion. Because of this we have to focus on the final mass, the mass at the end of the collapse.

Calculations have shown that white dwarfs occur in the collapse of stars that have a mass less than 1.4 solar masses. They are created in a slow collapse that occurs over millions of years. It isn't a dramatic ending; they just get smaller and smaller until they are perhaps 20,000 miles across, and at the same time they get denser and denser.

The formation of neutron stars, on the other hand, is dramatic. They arise in the collapse of massive stars, stars of at least 4 solar masses, stars that supernova. The supernova explosion rips most of the outer layers from these stars; approximately 80% of the mass is blown off. But in the process, the core is compressed. The electrons and protons of the core are, in effect, squeezed into one another, creating neutrons. The end result is a neutron star about 10 or 12 miles across.

But, again, in the case of neutron stars there is a limit. The final mass of neutron stars has to be less than 3.2 solar masses. What happens if the mass is greater than this?

BLACK HOLES

After Oppenheimer had finished his work with Volkoff on neutron stars, he turned his attention to stars more massive than 3.2 solar masses. He began his new study with a student named Hartland Snyder and the results they obtained in late 1939 completely surprised and stunned astronomers. Many, in fact, refused to believe them.

Upon examining these larger stars they found that stars more massive than 3.2 solar masses literally collapsed forever. There was nothing to stop them. In their words:

> When all the thermonuclear sources of energy are exhausted a sufficiently heavy star will collapse. The contraction will continue indefinitely . . . the radius of the star approaching [closer and closer to] its gravitational radius.

It's interesting that at this stage little was known about the gravitational radius. It was still considered to be a kind of singularity. Lemaître's work showing that it wasn't a singularity had appeared a few years earlier but it was still not widely known. It is possible that Oppenheimer learned of it from Robertson, but in his paper he said little about how the collapse of the matter of the star (the matter collapses in a short time to zero radius inside the gravitational radius) was to be reconciled with the view of the collapse as seen by an outside observer. To the external, distant observer, the star appears to stop collapsing to a finite radius, the gravitational radius.

When the Oppenheimer–Snyder paper was published, World War II had already started and Oppenheimer was soon off working on the Manhattan Project. He never returned to the problem, nor did his students. During the 1940s and 1950s, there was little interest in gravitational collapse or neutron stars. Zwicky continued working on neutron stars and whenever possible he made his views known. But no one paid any attention to him.

Even with the ending of the war in 1945 little attention was paid to the strange new stars. It was not until ten years later that the Princeton physicist John Wheeler, who had been working in nuclear physics, began wondering what the final state of the system of nucleons (e.g., protons or neutrons) would be when the number of nucleons was extremely large. Though his initial interest was not in astrophysics, he soon discovered that the problem took him deeply into the subject. Working with students Kent Harrison and Masami Wakano, he wrote a large computer program that would predict the end state for stars of different masses, under many different conditions.

The computer plots gave amazing results: they verified both the Chandrasekhar and Oppenheimer–Volkoff limits for white dwarfs and neutron stars. Wheeler and his students also looked into the possibility that stellar explosions or high rotational rates could change the end state of the star. They found that there was no escape: neutron stars *had* to exist. Furthermore, the strange end state of stars more massive than 3.2 solar masses, namely, indefinite collapse, found by Oppenheimer and Snyder was verified. But Wheeler was not satisfied with this result. It continued to bother him for years; he was sure there was a way around it.

Still, even with Wheeler's impressive results there was little interest in the topic in America. Then, in the early 1960s, astronomers discovered quasars: strange, energetic, enigmatic radio sources that appeared to be located beyond the most distant galaxies. Nothing this strong had ever been seen; they were incredibly energetic. What was supplying their energy? It was impossible that ordinary nuclear energy was responsible. Something else, something much more powerful was needed, and a startling suggestion was made within months of the discovery. In 1963, Fred Hoyle of Cambridge and William Fowler of Caltech published a paper speculating that quasars were supergiant stars in the process of collapsing to black holes. According to their calculations, these stars would emit tremendous amounts of energy as they collapsed.

Astronomers finally began to take an interest. Neutron stars, black holes, and quasars were mysteries demanding attention. What was generating the energy of quasars? Were they associated with gravitational collapse? Did the collapse lead to neutron stars or black holes? A conference was called to study the problems.

Now referred to as the First Texas Symposium on Relativistic Astrophysics, it took place from December 16 to 18, 1963 in Dallas, Texas. Over 300 scientists were invited. Excitement mounted as the participants assembled. One of the longest reports was given by Wheeler and his students, but there were other important new results. Jesse Greenstein of Mount Wilson Observatory talked about quasars. Roy Kerr of New Zealand reported on a new

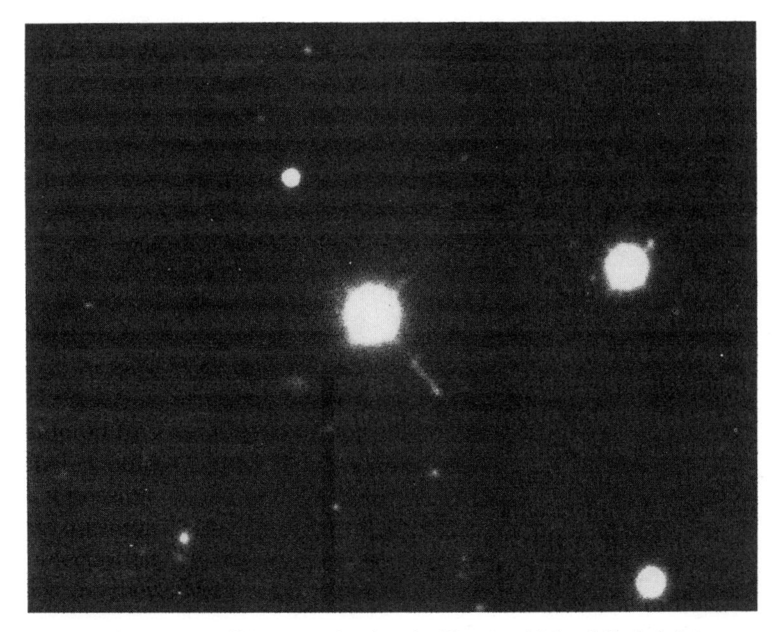

A quasar showing a small jet emanating from it. (Courtesy National Optical Astronomy Observatories.)

solution to Einstein's equations, one corresponding to the collapse of a spinning star.

But there was still no observational evidence for either neutron stars or black holes. The foundations, however, had been laid, and many people were now working in the area. Then, in 1967, news came that an important discovery had been made; Antony Hewish of Cambridge University had discovered a rapidly pulsing star (now called a pulsar). Others were soon found. Within a year Thomas Gold of Cornell had shown that they were neutron stars. Several hundred of these objects are now known to exist.

With neutron stars in their pocket, astronomers could now turn their efforts to black holes. Was there any observational evidence for such bizarre objects? Before we look into this, let's examine the properties of black holes as predicted by general relativity.

PROPERTIES OF BLACK HOLES

Black holes arise in the collapse of a massive star. Actually, the final mass of the star need only be greater than 3.2 solar masses, and, since many stars of this mass exist, it seems reasonable to assume that black holes should be quite common. As in the case of the neutron star, though, we have to be careful. During its life and in its final collapse, a massive star is likely to lose considerable mass. If a star is to survive with 3.2 solar masses, it probably has to be about 8 solar masses before the collapse.

It might seem as if a black hole is just like a neutron star, but slightly smaller. This isn't so; black holes are significantly different from both neutron stars and white dwarfs. White dwarfs are gaseous, just as stars like our sun are, but the gas is much denser. Neutron stars, on the other hand, are at least partially solid; they are believed to have a solid surface. We have, in fact, detected the solid shell-like surface of neutron stars cracking in a starquake.

Black holes are different from both of these objects in that they are basically empty space. All of the matter of a black hole is at the center in the form of a singularity—a point of infinite density and zero dimensions.

It's helpful to try to visualize a giant star collapsing to a black hole. It's unlikely we'll ever actually see this happen; giant stars collapse too fast. In a tiny fraction of a second it's all over, and even if you could catch it with a high-speed camera, the final stages would probably be obliterated by clouds. We'll suppose, however, that we can slow the process down and watch it in detail.

Initially the star is a giant. It's nuclear furnace flickers and gravity starts to overcome the star. Slowly it begins to collapse, but as the collapse proceeds it speeds up, just as a ball rolling down an incline gains speed. As it grows smaller it changes color, quickly moving through a rainbow of colors. Then it reddens, and as it

What a black hole might look like if you encountered one in space. You would be able to see it only because it blocks off background stars. (Courtesy Lick Observatory.)

does, the collapse slows down. Finally, it slows almost to a stop. At this stage we see a black sphere in space surrounded by large numbers of stars. Gravity is strong near the black hole and the space around it is severely curved.

The black sphere we see is called the "event horizon." It has a radius equal to the gravitational radius. Although it looks like a solid surface from a distance, it isn't. It's more like a black balloon, but, strangely, you could pass right through this balloon without popping it. Furthermore, it's only a few miles across.

The event horizon is appropriately named, for it is a horizon to our world. Once you pass through it you can never return to our world again, at least not in the same region of space. It is a one-way surface. If you were inside and trying to get out, you would find that the event horizon was receding from you at the speed of light. And, since you can't travel at the speed of light, you could never reach it.

Something else is strange. Space and time have interchanged their roles: where you once had control over space, you no longer do. You are pulled relentlessly into the singularity. But unlike the case on Earth, you now have control over time. Still, there is no escape; you will soon end up in the singularity.

Although you can't see it directly, the space around the black hole is warped into the shape of a wormhole. The black hole itself is at the narrowest part of the wormhole. This is, of course, the wormhole we would like to use to travel to the stars.

If you probe the region around the black hole with a powerful searchlight, you will see evidence of the tremendous warping of space. When the beam passes far from the black hole, say 50 times the gravitational radius, you will find that it is bent only slightly as it moves past the black hole. The space here is relatively flat. Turning it closer and closer to the black hole you'll find that the beam bends more and more. Finally, at 1.5 times the gravitational radius, the light takes on a circular path around the black hole. This is the photon sphere. Inside the photon sphere the beam will be pulled into the black hole.

Although a black hole is certainly bizarre in most respects,

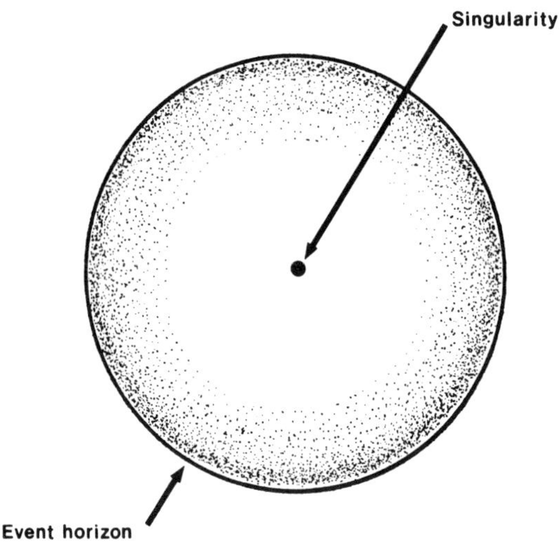

Structure of a black hole. The surface is known as the event horizon. Singularity is at the center. All the mass of the collapsed star is in the singularity.

there are ways in which it is not different from ordinary stars. If our sun collapsed and became a black hole overnight (of course, it won't because it's not massive enough) the Earth would continue in its orbit in the usual way. But the next morning we would notice the difference: it would seem as if the sun had disappeared, and we would remain in darkness. If you then jumped into a rocketship and followed the gravitational field down to where you could see the sun, you would see that it is now a small black sphere in space, visible only because it blocks off background stars. At this point you would find yourself in an extremely strong gravitational field; you are only a few hundred miles from the black hole and it is producing the same gravitational field as our sun did. When you are this close to it, the field is incredibly strong. Unless you keep your spaceship moving at high speed you will be pulled into the black hole.

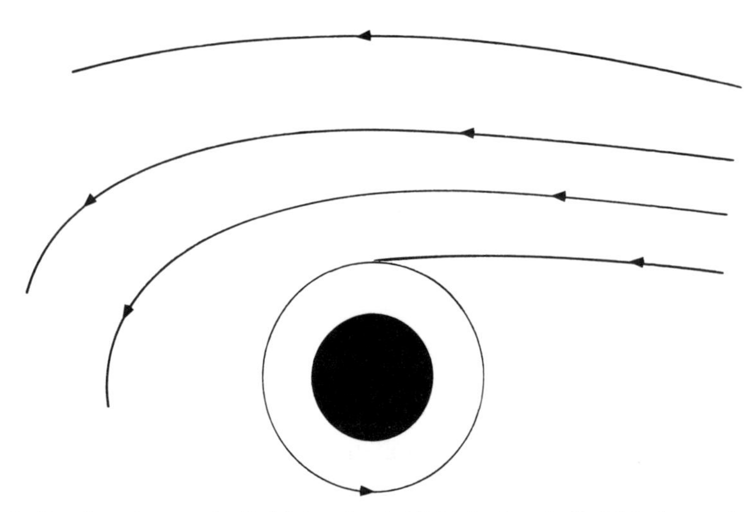

Probing the region around a black hole with a flashlight. Far from the black hole the space is only slightly bent and the beam is only slightly deflected. Closer to the black hole the space is more curved and the beam is bent more. At the photon sphere it goes in a circle around the black hole.

As we circle the black hole, let's assume we release a probe and let it fall into the black hole. Assume that the probe has a clock aboard that we can observe. We see the probe accelerate toward the black hole, getting smaller and smaller as it moves away from us. Soon, though, instead of accelerating, the probe begins to slow down. At the same time the clock aboard it begins to run slow, as compared to our clock. As we continue to watch it through our telescope, it moves slower and slower and its clock runs ever slower. Strangely, the probe never seems to quite make it to the surface of the black hole; furthermore, its clock almost stops, yet it never completely stops.

Suppose now that we put an observer aboard the probe (obviously a very brave or foolhardy one). Oddly, he sees a very different picture: within a very short time he passes directly through the event horizon into the black hole. To him, his clock is not slowing down; it is running normally. It is the clock back on the spaceship that is running strangely; it runs faster and faster as he gets increasingly closer to the black hole.

It's obvious from this that time behaves quite differently around a black hole. We see that there are two different views of how time passes: the point of view of the distant observer and that of the observer falling into the black hole. It is convenient therefore to label these times differently; we call the time as seen by the outside observer, coordinate time, and the time seen by the observer falling into the black hole, proper time.

But time is not the only thing that is affected as our observer approaches the black hole. As I mentioned earlier, the gravitational pull this close to a black hole is extremely intense; furthermore, it differs significantly over small distances. This means that the front of his probe will be pulled toward the black hole with a much greater force than the back end, and as a result it will be pulled apart. In short, it will be stretched as he gets closer and closer to the black hole, and of course his body will get stretched with it. The force that is responsible for this is called a tidal force.

Is there any way we can get around this problem? Indeed, there is. The magnitude of these tidal forces depend on the mass of the black hole. If it is sufficiently massive, on the order of thousands of solar masses, the tidal forces, even at the event horizon, are not large. You would pass right through the event horizon without any discomfort. There are, of course, other problems that we would also have to contend with, but we'll leave them for later.

GEOMETRY OF A SCHWARZSCHILD BLACK HOLE

The black hole that we have been dealing with so far is perfectly spherical and nonspinning. We refer to it as a Schwarzschild black hole. If we are interested in the possibility of using this black hole as a time tunnel to distant parts of the universe, we must look carefully at the geometry of the curved space around it.

To do this we have to use space-time diagrams. A space-time diagram is nothing more than a plot of time along one axis and space along the other; this single space axis, however, represents all three dimensions of space. (We talked briefly about such diagrams in Chapter 2.) We can obviously use a diagram of this type to plot a trip to a star, say, our nearest star, Alpha Centauri. We will label the vertical axis as the time axis and scale it in years. The horizontal axis is then the space axis, and we will scale it in light-years. Furthermore, we will assume that a line at 45 degrees to these axes (between them) represents the speed of light.

Note, first, that the line between Earth and Alpha Centauri can at no point have a slope of less than 45 degrees from the space axis, as this represents speeds greater than the speed of light. Also, since Alpha Centauri is 4.3 light-years away, this is the

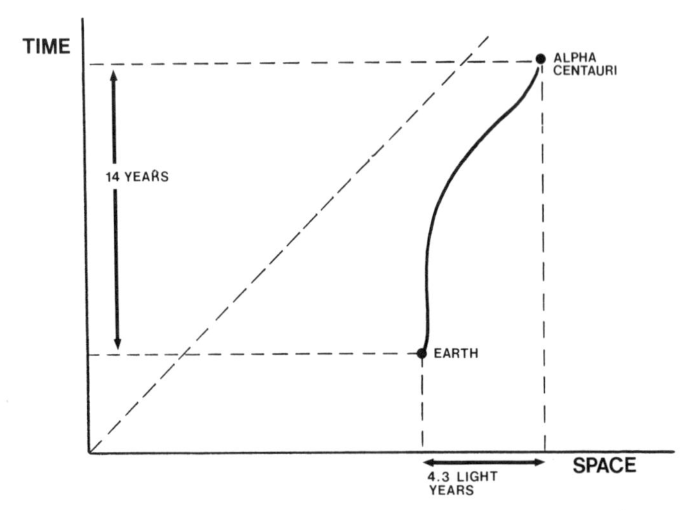

A simple representation on a space-time diagram of a trip to Alpha Centauri. The trip takes 14 years; the distance to Alpha Centauri is 4.3 light-years.

distance we will have to travel to get to it, as shown on the space axis. Assume that it takes 14 years; we represent this on the time axis.

The trip to Alpha Centauri, for the most part, is through flat space. But what would a trip through curved space be like, say between two points in the vicinity of a black hole. To illustrate it we need a space-time diagram around the black hole. As we saw earlier, a black hole consists of an event horizon at the gravitational radius and a singularity at the center. We can represent the region in and around the black hole as in the figure below. The singularity is shown as a jiggly line and the event horizon as a dotted line. Space is to the right and time is upward.

To illustrate a trip in this diagram let's go back to the spaceship we talked about earlier with the small probe attached to it.

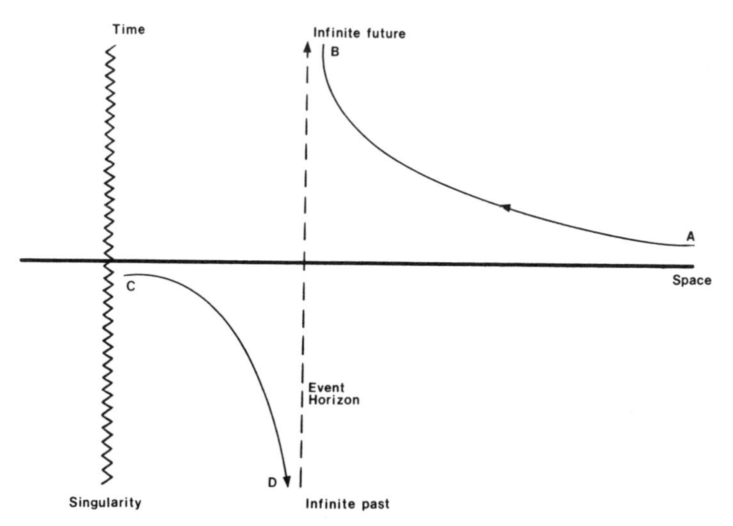

Simplest space-time diagram for the space around a black hole. A traveler starting at A and moving toward the event horizon never reaches it. This is represented by the line AB. Note that it never quite touches the dotted line. The trip CD is inside the event horizon. A traveler trying to get out never quite reaches the event horizon on the inside. These "infinities" are a problem. Because of them scientists prefer to use Kruskal's diagram (see next figure).

Again, we'll stay safely in the spaceship and let the probe examine the black hole. We assume that we have a method of observing the clock aboard the probe. As the probe moves toward the black hole we see that its clock slows down compared to ours. As it gets closer it slows down more and more. We can illustrate this approach to the event horizon as the line from A to B in our diagram. Note that this line gets closer and closer to the event horizon (the dotted line) but it never quite reaches it.

We know, however, that if an observer were in the probe, he would see things quite differently. Within a short period of time he would pass through the event horizon into the black hole. Let's assume, then, that our observer is inside the event horizon, say, near the singularity, and he tries to get out. He will move backward in time (coordinate time), approaching the event horizon from the inside. He will approach the event horizon closer and closer as he goes back further and further into the past. But he will never quite reach it. His path is shown as CD in the diagram.

The complexities, particularly the infinities, of this diagram are obviously difficult to deal with. We talked briefly about them earlier and mentioned that Kruskal devised a different and better diagram. In Kruskal's diagram, the upper half of the line representing the singularity is moved to the top of the diagram and the lower half is moved to the bottom. Note, however, that time is still plotted vertically and space horizontally. The two lines at 45 degrees represent the event horizon.

Again, as in the previous diagram, we can represent various trips in and around black holes. Three such trips are shown; they are labeled A, B, and C. Trip A is between two points outside the event horizon. We are outside the black hole, so except for the fact that the space is curved, this is the same as any trip between two points in space.

Trip B, however, is quite different. This is a trip through the event horizon and into the singularity. Note that in this diagram we do not have the trajectory approaching the event horizon asymtotically (increasingly close) as we did in the previous spacetime diagram. The most interesting trip of the three, though, is C.

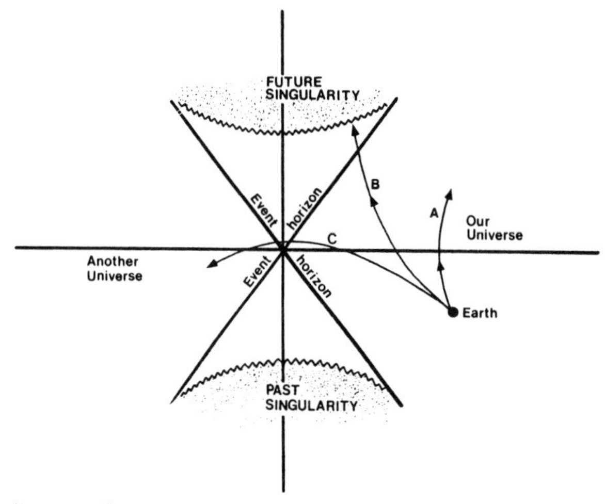

Kruskal's diagram. There are now two event horizons, and they are at 45 degrees to one another. Singularities are shown at the top and bottom. Three possible trips in and around the black hole are shown (A, B, and C).

In this case we pass through two event horizons and then into another universe. In particular, we avoid the singularity. This trip corresponds to entering the space-time tunnel or wormhole at one end, moving through it avoiding the singularity, and then passing out through the other end to another universe.

If you look carefully, though, you see a problem. I mentioned earlier that we cannot make a trip where the trajectory has an angle less than 45 degrees; this corresponds to a speed greater than that of light. Trip C requires such a trajectory and is therefore impossible. This tells us that the wormhole associated with a Schwarzschild black hole is of little use to us; we can't use it to travel to another universe or to a distant point in our universe.

There is, however, another possibility. Most stars spin and

when they collapse they create spinning black holes. Einstein struggled for years to find the solution to his equations for a spinning black hole, but the difficulties were formidable and he was never able to overcome them. But would the wormhole associated with a spinning black hole allow us to pass through? We will look into this possibility in the next chapter.

CHAPTER 8

Spinning Gateways

We saw in the preceding chapter that Schwarzschild black holes have space-time tunnels, but they are of little use to us because a speed greater than that of light is needed to get through them. All hope isn't lost, however, as there are still spinning black holes. The solution for this type of black hole was found by the New Zealand mathematician Roy Kerr in 1963, while he was at the University of Texas. Kerr showed that a spinning black hole was quite different from a nonspinning one. It had an event horizon, but there was another surface outside the event horizon that touched it at the spin axes and had maximum separation at the equator (see figures). We refer to this surface as the static limit. The region between it and the event horizon is called the ergosphere.

Once a solution was at hand scientists turned their attention to the space-time tunnel. Could we get through it with a speed less than that of light? When the calculations were done, they found that it was possible.

But was this type of black hole likely to exist? It turns out that literally all stars spin; our sun, for example, rotates on its axis in roughly a month. Furthermore, even if a star only has a small spin initially, when it collapses and becomes a black hole, it will end up with a high spin. The reason for this is the conservation of angular momentum (spin) which says that the total angular momentum before the collapse has to be equal to the total angular momentum after. It tells us that as the radius of the star decreases, its spin increases. You are likely familiar with this phenomenon in relation

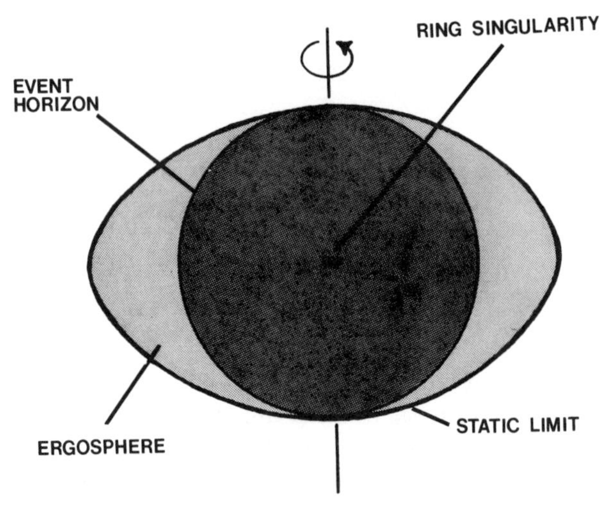

Side view of a Kerr black hole. The event horizon is still spherical, but there is another surface around it called the static limit. Note that the two surfaces touch along the spin axis. The region between the two surfaces is called the ergosphere. Note that the singularity is now a ring. The event horizon is actually double in most cases, but this is not shown in this diagram (see next figure).

to figure skaters. They begin a spin with their arms outstretched, then pull them in; as they do this they spin much more rapidly. In the same way when a star pulls its matter in (i.e., collapses) it also spins faster.

So it was possible, at least in theory, to get through the tunnel associated with the Kerr black hole. But as the details were worked out it became clear that black holes were more complex than anyone had imagined. A tremendous effort would be needed to thoroughly understand them. But considerable interest had been generated and soon many people were working on black holes. Within a few years several important breakthroughs were made.

During the early and mid 1960s the Russians remained at the forefront of black hole physics. Two important papers were published in the mid 1960s by Yakov Zel'dovich and I. D. Novikov that

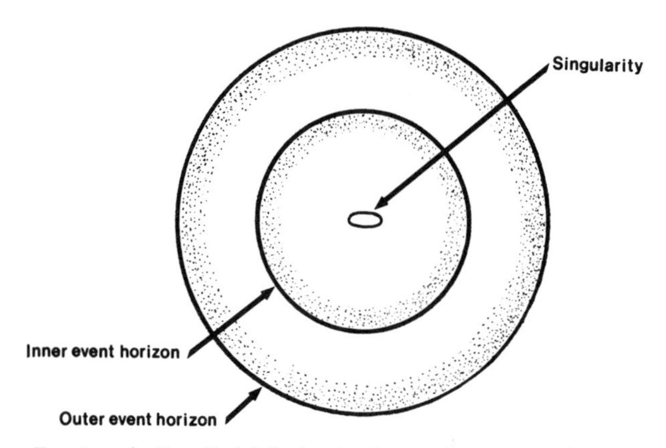

Top view of a Kerr black hole showing inner and outer event horizons.

laid the foundations of relativistic astrophysics. They gave a detailed discussion of the Oppenheimer–Snyder paper which left no doubt about its correctness. They also talked about the appearance of the black hole as seen by a distant observer, and speculated as to what went on inside the event horizon. But they emphasized that the interior of the hole was forever cut off from us.

THE "NO HAIR" THEOREM

A particularly important breakthrough of this era was made by Roger Penrose. Son of a well-known biologist and brother of a British chess champion, Penrose loved to work on difficult problems, and even as a youth spent considerable time devising challenging games and puzzles. He started out as a mathematician, but about 1960 he switched to general relativity and was soon working on black holes. In 1965, while at Cambridge, he proved a theorem showing that a singularity had to develop when an event horizon formed.

Roger Penrose.

Penrose's work eventually came to the attention of the Canadian physicist Werner Israel. At this stage there was still considerable confusion about the end state of the collapse of a star and the type of black hole that would be formed. It seemed as if there should be many different types of black holes. After all, stars are all different (have different properties), and since black holes are formed when stars collapse, black holes should also be different.

Israel showed that this is not the case. He proved that for a nonspinning black hole the only important property is mass. All black holes of the same mass are identical. It didn't matter what went into the black hole; the only thing left after it formed was a spherical event horizon with a singularity at the center. And the

Werner Israel. (Courtesy Lotus Studios.)

size of the event horizon depended only on the mass of the star that collapsed to create it.

For a while, Israel's discovery caused considerable confusion. According to his calculations, the end state of the collapse, the event horizon, had to be perfectly spherical. But it was unlikely that a collapsing star would remain spherical throughout its collapse; it would probably oscillate to some degree and this would cause a deviation from sphericity. Could such an object end as a black hole? A number of people speculated that it might end with no event horizon, as a naked singularity (singularity with no event horizon around it). The confusion was finally cleared up, however, when it was shown that any out-of-roundness would disappear just before the black hole formed and the event horizon would be perfectly spherical.

Israel's discovery was soon referred to as the "no hair" theorem (making a pun of the fact that black holes could not have hair, only mass). It was extended to rotating or Kerr black holes by Brandon Carter in 1970. He showed that Kerr black holes could only have mass and spin; for a given mass and spin they were all identical.

PRIMORDIAL BLACK HOLES

Even though there was now a way through Kerr tunnels, there was still the problem of tidal forces. (We talked about them earlier.) They are forces that tend to pull you apart when you come close to a black hole. They are weak only if the black hole is extremely massive and stellar black holes are not massive enough. In 1970, however, Stephen Hawking of Cambridge University showed that another type of black hole might exist; this type, according to his calculations, would have arisen in the big bang explosion that created the universe.

Born in 1942, Hawking was the eldest of four children. At the age of 14 he had already decided to become a physicist. His father was sure, though, that he would never be able to find a job in such a profession and tried to direct his interests elsewhere. But Stephen had made up his mind. Upon completing his undergraduate work at Oxford, he went on to Cambridge for graduate studies. It was during this time that he developed the first symptoms of amyotrophic lateral sclerosis, a crippling neurological disease. Within a short time he was restricted to a wheel chair and could hardly speak. But his determination was strong and he was soon making important contributions to black hole physics.

Stephen Hawking.

In the late 1960s, Hawking began to look at the details of the big bang explosion. He realized that if it had been perfectly uniform, in other words, if the gas had expanded out uniformly in all directions, the universe would have ended as an infinitely thin uniform gas. But it was obvious that this didn't happen. Furthermore, explosions here on Earth do not expand outward uniformly; small overdense regions develop. Hawking was sure that the same thing happened in the case of the big bang, and, if it did, some of these regions would have been compressed into black holes. These black holes would be quite different from stellar collapse black holes. There would be a large range of masses, from tiny ones, about the size of an atom, up to extremely massive ones, millions of times more massive than our sun. Some of these massive black holes may, in fact, be at the centers of galaxies. We usually refer to them as primordial black holes to distinguish them from the stellar collapse variety.

But do we have any evidence that such black holes actually formed? We do, and it's right in front of our noses. I'm referring to the galaxies. If the big bang explosion had been perfectly uniform, galaxies would not have formed. But we know they did and this means that the explosion must not have expanded outward uniformly.

We also have indirect evidence for massive black holes in the form of active galaxies. Some galaxies appear to be undergoing tremendous explosions, and according to most astronomers the only way to explain their energy is to assume that there are huge black holes at their core.

I also mentioned that besides massive black holes, tiny ones, sometimes called mini black holes, would have formed. They are intriguing objects that have inspired considerable speculation. Some of them may have been as small as elementary particles. A number of people have suggested that some of them may have struck the Earth in the past. Others have suggested that the sun may have captured a number of them, and they may still reside in its core. We have no evidence that either of these is true, but they are interesting possibilities nevertheless.

EXPLODING BLACK HOLES

The early 1970s brought further developments in black hole physics. First came Penrose's discovery of a process that is now named after him. Riding a train into London one day, he began thinking about what he was going to present in his lecture that day. He thought about the ergosphere. A particle, or projectile, that entered this region could escape as long as it didn't pass inside the event horizon. He did some calculations, then began wondering what would eventually happen to a particle in the ergosphere. What if it broke up into two particles? He considered the possibility that one of the two particles fell into the black hole and the other escaped through the static limit. His calculations showed him that the particle that emerged would come out with more energy than the original particle entered with. But where was the extra energy coming from? A simple calculation showed him that it was coming from the black hole itself. Penrose was amazed; energy could be extracted from a black hole. It didn't seem possible.

One of those who became intrigued with the Penrose process was a 21-year-old Princeton graduate student by the name of Demetrios Christodoulou. Working on a thesis under John Wheeler, Christodoulou looked into the changes that a black hole undergoes as large numbers of particles entered it. He assumed that each particle broke up and one of the two resulting particles escaped, taking some of the energy of the black hole with it. How much energy could be extracted? Christodoulou found that there was an end state where no more energy could be extracted; he called it the "irreducible mass."

Hawking pondered Christodoulou's result. What did it mean? After considerable thought he was able to show that it was a special case of something much more general. The surface area of the black hole, the event horizon, could never decrease; it could only increase. This meant that if two black holes merged, the surface area of the overall black hole had to be greater than the combined areas of the individual ones.

It soon became obvious that if this was correct, the surface of the black hole was like entropy. Entropy, as we saw in an earlier chapter, is a measure of the disorder of a system. There appeared to be a distinct similarity between entropy and the surface area of a black hole. We know that the entropy of any isolated system always increases, just as the surface area of a black hole always increases. Similarly, if two objects of entropy A and entropy B come together, the resultant entropy is always greater than A + B, just as the combined surface areas of two black holes is always greater than the sum of their individual surfaces.

Hawking, together with Brandon Carter of Cambridge and

James Bardeen of Yale University in the United States, followed up on the similarity. They found that there was a complete analogy between black hole physics and thermodynamics (the study of heat). Just as there were three basic laws of thermodynamics, so too were there three laws of black hole physics. It was, in fact, amazing how similar the two sciences were. But Hawking cautioned that it was only an analogy and shouldn't be taken as anything more.

About this time, Wheeler, at Princeton University, called one of his graduate students, Jacob Bekenstein, into his office. Beckenstein was going to do a thesis on black holes. Wheeler explained to him that he was concerned about the concept of entropy in relation to black holes. The entropy of the universe always increases according to one of the basic laws of thermodynamics, he said. But if you took a system in which an entropy increase had just occurred and threw it into a black hole, evidence of the increase would be concealed. It would have no effect on the universe.

Wheeler told Bekenstein to think about this problem and see if he could come up with anything. A few days later Bekenstein came back with a solution. When you throw something into a black hole, he said, you increase its entropy, and since a black hole's surface area is a measure of entropy, it will increase this area. He said he believed that the surface area of a black hole was not just something that was analogous to entropy, but, apart from a constant, was a direct measure of it. But entropy depends on temperature. Bekenstein therefore needed something to represent temperature. Gravity, he was sure, would fit the bill.

But this introduced a problem. Black holes absorb everything; they are black and therefore have a surface temperature of absolute zero (lowest possible temperature in the universe). Because of this, Bekenstein had second thoughts about a relationship between temperature and gravity. He wrote, ". . . we emphasize that one should not regard [the surface gravity] as the temperature of the black hole: such an identification can easily lead to all sorts of paradoxes, and this is not useful."

Hawking heard about Bekenstein's work and became alarmed. He was convinced that there was an analogy between thermodynamics and black hole physics, but it was heresy, as far as he was concerned, to suggest that the surface area actually represented entropy. In 1972 Hawking wrote a paper with Carter and Bardeen that was, to some degree, directed toward Bekenstein. They emphasized that to push the analogy between thermodynamics and black hole physics too far was a mistake.

Then, in 1973, Hawking went to Moscow. He talked with Yakov Zel'dovich and Alexander Starobinsky about the problem. They told him about some work they had done indicating that rotating black holes might emit radiation and particles. Hawking was reluctant at first to accept the result, but they eventually convinced him. When he returned to England, he began looking into the mathematics of the problem. Sure that there was a better way to look at it, he decided to apply quantum theory to black holes, and what he found amazed him. He discovered that not only rotating, but also nonrotating black holes emitted particles and radiation. He was sure that he had made an error in his calculations. Furthermore, he was worried that if Bekenstein heard of the result, he would use it to further his argument for black hole entropy. He rechecked, but there seemed to be no error.

To understand Hawking's result we must begin with a prediction of quantum theory called virtual pair production. According to this prediction, pairs composed of a particle and an antiparticle (e.g., an electron and a positron) are spontaneously produced out of the vacuum. They pop into existence, then come back together almost immediately and annihilate one another. The process is allowed because of one of the basic principles of quantum theory, called the uncertainty principle. This principle tells us that there is a "fuzziness" associated with nature at the atomic level. The process is concealed within this fuzziness.

Hawking thought about the production of virtual pairs in the strong tidal forces just outside the event horizon. It was obvious that such pairs would be produced in abundance. In a sense they would be "ripped" out of the vacuum by the tidal forces. Further more, Penrose had shown that if one particle fell into the event horizon, the other would emerge with more energy than it came in with. Hawking therefore considered the possibility of one of the virtual particles falling through the event horizon and the other escaping. Using quantum theory he showed that the process was possible. Furthermore, some of the particles that emerged would produce radiation as they exited. From a distance, the black hole would appear to be emitting particles and radiation.

But if radiation was being emitted, the object had to be hot. Was Bekenstein right? Did the black hole actually have a temperature greater than zero degrees absolute? Hawking was sure that this was crazy, but if the black hole did radiate, it had to be hot. This meant that it wasn't a "true" black hole; black holes only absorb radiation. Furthermore, if radiation was being given off, energy was also being released. And if the black hole was losing energy, it was also losing mass, since Einstein's relation tells us that energy and mass are equivalent.

All of this seemed to imply that black holes evaporated. Hawking was sure that this was impossible. It was in direct conflict with the important theorem he had published only a few years earlier stating that the surface area of a black hole can only increase. Hawking realized, though, that this was a classical result (obtained using classical theory) and the prediction of evaporation was a quantum result. In this sense there was no conflict: both results could be correct within their sphere of influence. In the physics of ordinary large-scale objects we use classical theory, but when we are dealing with atoms we must use quantum theory. If we try to apply classical theory to atoms, we get the wrong answer.

But if black holes are hot, would we still be able to use their space-time tunnels for space travel? Hawking calculated the temperature for several cases and found that there was no problem. The only ones that were very hot were the mini black holes. A black hole with a mass equal to our sun would have a temperature so close to absolute zero that it could not be distinguished from it. A smaller black hole of mass 10^{25} grams, on the other hand, would have a temperature of 10 K (degrees absolute), and a tiny one, about the size of an electron, with a mass of 10^{15} grams, would have a temperature of 10^{11} K.

If the temperature of the tiny black holes is so high, we might wonder what the limit is. Obviously, the higher the temperature, the greater the radiation and the faster it evaporates. Tremendous amounts of energy and radiation are therefore lost in the final moments of a mini black hole's life. In effect, they explode. Calculations show that a mini black hole with a mass of 10^5 grams will last only 10^{-12} second. One of 10^{10} grams, on the other hand, will last three seconds and one of 10^{15} grams will last ten billion years. But ten billion years is the approximate age of our universe. This means that black holes that were formed with an initial mass of 10^{15} grams would be exploding right now, giving off an energy equal to that of about ten million 1-megaton nuclear bombs. If such a blast occurred in our solar system, we would certainly be able to detect it. Has it happened? As far as we know it hasn't, at least not in the last few hundred years.

Hawking was severely disturbed by this strange result. Black holes exploding! It seemed insane. He was worried that his reputation would be ruined if he reported it. Fellow scientists would think he had gone off the deep end. It had happened before with other well-known scientists. Toward the end of his life Eddington introduced his "fundamental theory," a theory that was never accepted and one that most scientists look upon today as a little crazy.

Hawking rechecked his calculations. Everything seemed to be correct; he could not see any holes in his reasoning. He talked to some of his close friends, Penrose, Sciama, and others about the result. They encouraged him to report it. So in the winter of 1974 he gave a talk at Rutherford–Appleton Labs in England. Many of the people in the audience were experimentalists, some were elementary particle physicists, and most found it difficult to follow the subtle mathematical details of his reasoning. A few, however, did understand them, and they shook their heads. It was impossible. As Hawking had expected there was a backlash against the discovery and several papers were soon published blasting it as irresponsible.

But as scientists began to look at the mathematics in detail, they soon began to realize that he was right. Furthermore, it appeared that the breakthrough was not just one in black hole physics, but in physics in general. Hawking showed that the radiation from black holes obeyed a basic quantum formula, called Planck's formula. This was the first time a link had ever been forged between classical theory (general relativity) and quantum theory.

DETAILS OF KERR BLACK HOLES

Earlier we saw that Kerr black holes are quite different from Schwarzschild black holes. Like Schwarzschild black holes they have a spherical event horizon, but around it they have a static limit, a surface that touches the event horizon at the spin axes and has maximum separation at the equator. The singularity of the Kerr black hole is also different; rather than a point, as in the case of the Schwarzschild black hole, it is a ring in the plane of the equator.

We saw that in the Schwarzschild black hole there was no escape from the singularity. If you fell through the event horizon, you would be sucked into the singularity. Space is infinitely curved here and you would be torn apart even before you entered it. But in the case of the Kerr black hole there is an escape. Only if you approached the ring edge-on, in the equatorial plane, would you be pulled into it. If you approached it from directly above or at some angle, for example, you could pass through its center. Once through it, though, you would be in for a surprise; on the other side you would encounter a world of antigravity. Instead of objects attracting one another, they would repel one another.

There are also other differences in the case of the Kerr black hole. The best way to illustrate them is to begin with a Schwarzschild black hole and gradually add spin. In the Schwarzschild black hole we have a point singularity and a spherical event horizon. As you add spin, something strange begins to happen.

The event horizon starts to shrink and at the same time an inner event horizon forms just above the singularity. The singularity also becomes ring-shaped. As you add more spin, the outer event horizon continues to move inward and the inner one moves outward.

Earlier, I mentioned that there was a static limit associated with the event horizon of a Kerr black hole, with a region called the ergosphere between them. In this case there is a static limit associated with each of the event horizons. As you continue adding spin, the two event horizons, along with their static limits and ergospheres, continue to approach one another. Eventually, when the spin rate is extremely high, the two event horizons

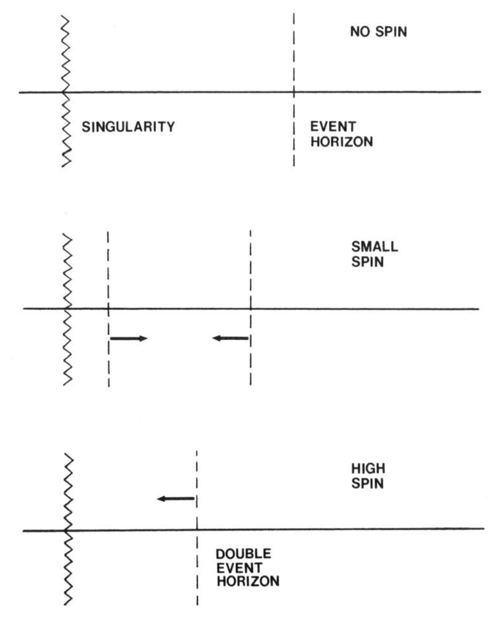

A simple plot of the singularity and event horizon showing how changes occur when you start with no spin and add spin.

merge to form a single event horizon. For obvious reasons this is called the extreme Kerr black hole.

What happens if we continue to increase the spin after a single event horizon appears? It turns out that the single event horizon moves inward, and if enough spin is added, it eventually disappears. We are then left with what is called a naked singularity (a singularity with no event horizon).

On the basis of this, we have to wonder what type of Kerr black hole would be formed in the collapse of a spinning star. In other words, how fast would it be rotating when it formed the black hole? It's important to remember that angular momentum is conserved, so even if the star was spinning relatively slowly, it would be spinning rapidly when it became a black hole. In 1974, Kip Thorne of Caltech looked into this. He showed that most stars that collapse to form Kerr black holes would be spinning at a tremendous rate, so fast that their two event horizons would be almost together.

IN ORBIT AROUND A KERR BLACK HOLE

If you approached a Kerr black hole in a spaceship, you would notice something strange. As a Kerr black hole spins, it drags the space close to it around with it, and if you entered this space, you would be pulled in the direction of spin. This is referred to as frame dragging.

A long way from the black hole, the effect is relatively weak and you can easily escape by firing a retro-rocket. As you get closer to the black hole, though, you will be pulled at high speed and it will become difficult to overcome the pull. Finally, once you are inside the static limit, and therefore inside the ergosphere, you can no longer remain at rest, regardless of how powerful your jets are. This is, in fact, where the name "static limit" comes from. You cannot remain static because it would require a speed greater than that of light in a direction opposite the spin of the black hole, and no one can travel that fast.

Although you cannot remain at rest, you are not trapped; you can still escape by pointing your spaceship outward. Not until you are inside the event horizon are you completely trapped. Interestingly, if your rocket entered in through the event horizon and exited out through the static limit, you would come out with a much greater speed than you entered with. You would, in fact, take some of the energy of the black hole with you.

Suppose now that you are just outside the black hole and have a large searchlight. How will the beam act if you direct it toward the black hole? In the case of the Schwarzschild black hole, we talked about the photon sphere, the surface around the black hole where a light beam went into a circular orbit. Does a Kerr black hole also have a photon sphere? Indeed, it does. If you start a Schwarzschild black hole spinning, you find that just as the event horizon breaks in two, so too does the static limit. One moves outward from the position of the Schwarzschild photon sphere and the other moves inward.

If you are in the plane of the equator and send a beam of light tangentially toward the outer photon sphere, you will find that the beam will get caught up in it only if it is directed opposite the spin of the black hole. We refer to this as the counterrotating photon

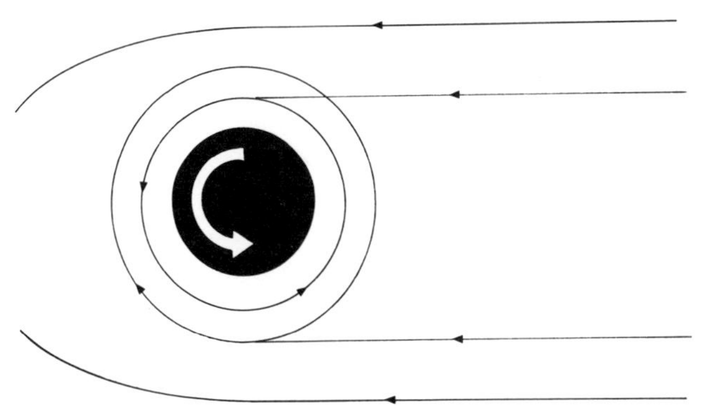

Probing the region around a Kerr black hole with a flashlight. There is an inner and an outer photon sphere.

sphere. If you then switch the beam over so it is orbiting in the same direction the black hole is spinning, you find you can probe much closer to the black hole before the beam gets pulled into a circular orbit in the inner photon sphere. This is the corotating photon sphere.

A TRIP INTO A KERR BLACK HOLE

Let's turn now to a trip into a Kerr black hole. As in the case of a nonspinning black hole, we will need a space-time diagram. In the previous case we used the Kruskal diagram. In this case, however, it will be more convenient to use what is known as a Penrose diagram. Looking back at the Kruskal diagram, we see a problem. Occasionally we want to talk about what happens at an infinite distance from the event horizon, but infinity is not in our diagram. Penrose showed, however, that it could be brought into the diagram in a simple way. There are, of course, several infinities that have to be brought in: future and past infinities for both our universe and the "other" universe. Penrose brought them in as two

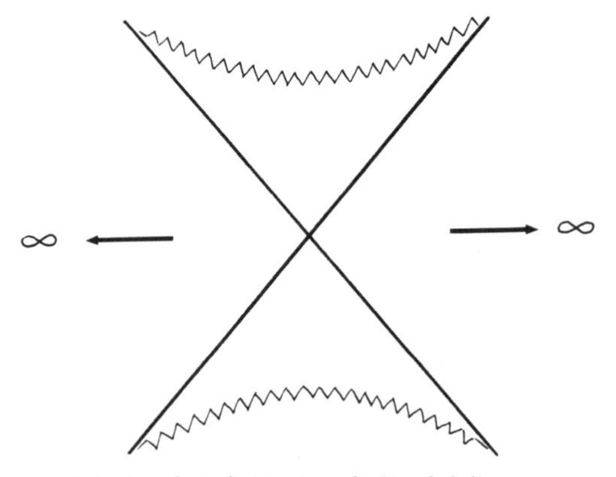

Bringing the infinities into the Kruskal diagram.

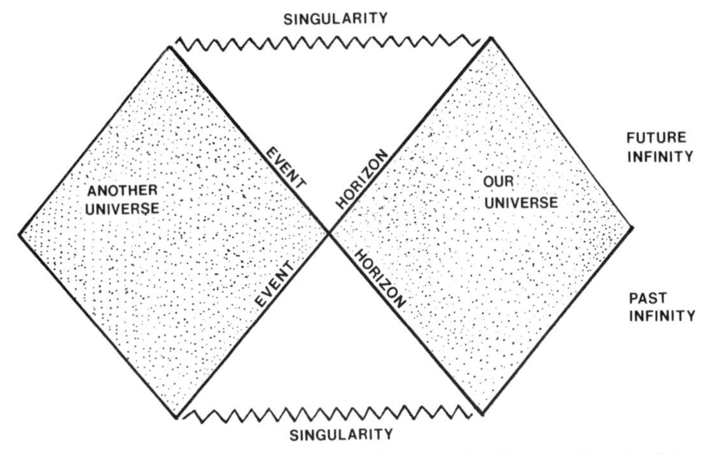

A Penrose diagram. All four infinities are shown in the diagram. For simplicity the singularity is shown as a straight line rather than a curved one.

angled lines (at 45 degrees) on the left and right of the diagram (see figure). But a cursory glance at this diagram shows that there is a difficulty. If we try to pass through the event horizon to get to the universe on the other side, we cannot. It requires a speed greater than that of light. Yet earlier I mentioned that we could use Kerr black holes to get to other universes. What is the problem? The problem is that the diagram as shown here is incomplete. The entire Penrose diagram for a Kerr black hole is quite complicated; it is shown in the following figure (note that the above diagram has been turned sideways; it is seen in the center part of this diagram).

Again, as in the case of the Schwarzschild black hole we can consider trips in and around Kerr black holes. Forty-five degrees still represents the speed of light, and we are, of course, unable to travel that fast. Our trajectory therefore cannot, at any point, be greater than 45 degrees from the vertical.

Let's begin with trip A shown in the diagram. In this trip, we pass through the outer event horizon; as we do so, space and time reverse their roles. Continuing on, we come to the inner event

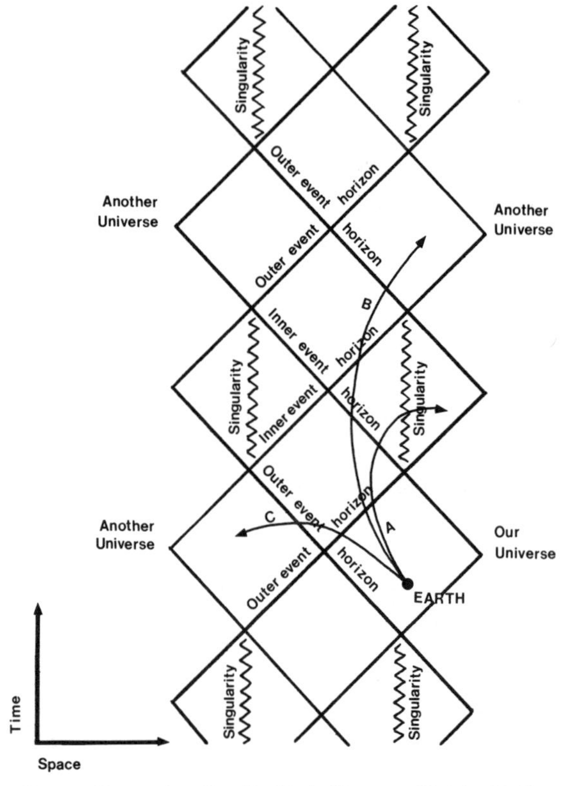

The complete Penrose diagram for a Kerr black hole. Three possible trips (A, B, and C) into a Kerr black hole are shown.

horizon; passing through it, we get another reversal of space and time, bringing us back to where we were when we left. We are now inside the region between the two event horizons and the singularity. The singularity is, of course, a ring and if we approach it out of the horizontal, we can pass through it into an antigravity universe. This is the case depicted in the diagram.

Trip B is perhaps of more interest, since we will also end up in another universe, but gravity will be the same as it is in ours. Again we pass through an outer and inner event horizon. Then we pass through a second pair of outer and inner event horizons and beyond the inner one we enter another universe.

The concept "other universe," as you might expect, bothers many astronomers, as it is difficult to visualize what it really means. It turns out, however, that the "other universe" can equally well be taken to mean a distant point in our own universe, distant from us both in space and time. This means that if we could overcome the problems, we may be able to use Kerr black holes to travel to distant points in our universe.

Finally, getting back to trip C, we find that it involves a path with an angle greater than 45 degrees and is therefore impossible. This means that we can't reach the universe directly across from us.

OTHER TYPES OF BLACK HOLES

So far we have encountered two types of black holes: nonspinning and spinning. The nonspinning variety occurs because mass is preserved in the creation of a black hole, or stated another way, because the gravitational field produced by the mass is able to escape through the event horizon. We have also seen that frame dragging is preserved in a Kerr black hole. A collapsing star causes frame dragging, and when the star becomes a black hole, frame dragging continues. Because of this we can distinguish a Kerr black hole from a nonspinning one.

What about other types of black holes? If the gravitational field reaches through the event horizon, we might ask if there are other types of fields that do the same thing. And indeed there are: the electromagnetic field. A black hole with a charge could be distinguished from the other two types as a result of its electric field. The mathematical solution for such a black hole was discovered in 1917 by Reissner and Nordström, and this type of black hole is now named after them. In theory it is an important type, but in practice it seems unlikely that a black hole with charge would exist. The problem is that very few stars have an excess charge on them. Our sun, for example, has no charge. If a star develops an excess charge, it is usually soon neutralized.

Is there any other property that can distinguish a black hole? The answer is no. As far as we know, only mass, spin, and electrical charge are preserved in the creation of a black hole. This was the essence of the "no hair" theorem we talked about earlier. Israel first proved it for mass alone, but it was later extended to spin and charge.

Looking at these three properties, though, we can see that another type of black hole is possible: one that has both spin and charge. The solution to Einstein's equations for this case was found by G. T. Newman of the University of Pittsbugh and several of his students, so it is now referred to as the Kerr–Newman black hole.

In all, then, there are four possible types of black holes. Each is of interest theoretically, but the only one likely to occur in nature is the Kerr black hole.

There is, however, something we have been ignoring. At the bottom of all the space-time diagrams we have been using is a region of reversed time. This region suggests that a time-reversed black hole might exist. Furthermore, if we look at a diagram of a space-time tunnel, it is obvious that only one end of it can correspond to the black hole. A black hole is a region where matter is pulled in. If matter is pulled in at one end of the space-time tunnel and somehow avoids being crushed in the singularity, it must exit at the other end. In effect, it must be ejected at the other end, which means that the other end must be a "gusher." Astronomers refer to this end as a white hole.

But do white holes exist? It turns out that mathematically they pose many problems. Theory predicts that they had to have existed for all time—or at least since the big bang. This means that they can't be associated with stellar collapse black holes. They can only be associated with black holes that were created in the big bang: primordial black holes. But this introduces another problem.

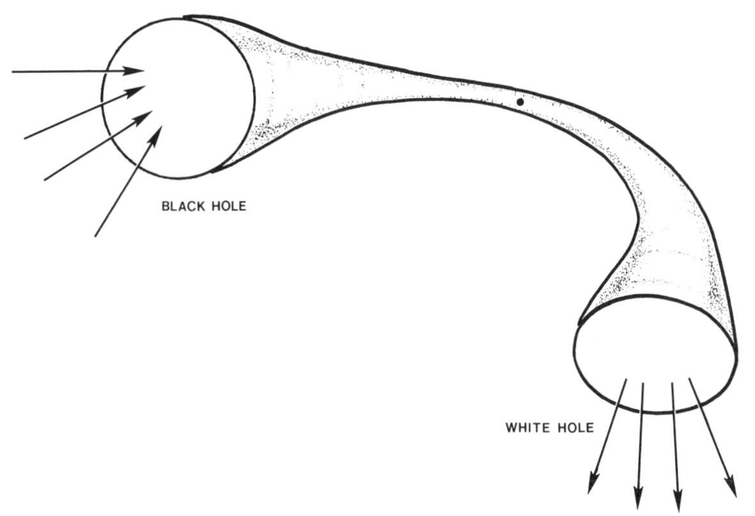

Space-time tunnel of a black and white hole. The singularity of the black hole is shown as a dot at the center. One end is a black hole, the other end is a white hole.

When first created, all primordial black holes were extremely small, since they were created about 10^{-20} second after the big bang; at that stage, the universe itself was small. Earlier, though, we saw that tiny black holes emit particles and radiation at a prodigious rate. But white holes also emit particles and radiation at a high rate. How would you distinguish tiny white holes from tiny black holes? Stephen Hawking has shown that you can't; they are indistinguishable. Furthermore, Douglas Eardley of Yale University has shown that even if white holes attempted to form in the early universe, black holes would quickly form around them and we would end up with only black holes. White holes, it seems, might not exist. But this would mean that our space-time tunnels would have entrances, but no exits.

So we have problems—many problems. Exits may not exist, and even if they do, space-time tunnels may pulse so fast that we wouldn't be able to get through them. In addition, we can only get through those associated with extremely massive black holes because of the tidal forces.

Furthermore, we still have to ask ourselves: Do black holes really exist? Everything we have said so far is based on theory—Einstein's general theory of relativity. Do we have any direct observational evidence for black holes? We will look into that question in the next chapter.

CHAPTER 9

Searching for Time Tunnels

Theoretical evidence for black holes, no matter how good, is of little use if we don't have any observational evidence to back it up. How, then, do we go about getting observational evidence? It obviously isn't easy. Black holes are only a few miles across and emit no light. Furthermore, they are black, the same color as the background sky. If we were close enough to one, we would be able to detect it indirectly, since it would block off the light from background stars and would appear as a black sphere. But from Earth we wouldn't be able to see this sphere. The nearest black hole is probably light years away, and even with our largest telescopes we couldn't see anything that small.

It would seem, on the basis of this, that there is no way we are ever going to detect a black hole by looking for one directly. Strangely enough, there is one possible way. We saw earlier that according to Einstein's general theory of relativity, a light beam is deflected as it passes a gravitating object. In 1936, Einstein showed that gravitating objects could be used to focus and perhaps magnify the image of distant objects. If a gravitating object, say a black hole, was directly in line with a distant galaxy, the light from the galaxy would bend around the black hole as it passed, and the galaxy would appear to us to be a ring of light. Such rings, or at least parts of rings, have been discovered and are now referred to as Einstein rings. Furthermore, if the alignment is not exact, rather than a ring we would obtain two or more images of the galaxy.

The only problem with this method is that you don't need a

black hole to get an Einstein ring, or even two or three images. Any dense object will do. And all of the gravitational lenses, as they are called, that have been discovered so far are believed to be due to galaxies. It is possible, though, that one day we will discover one that contains a black hole.

What about other ways of detecting black holes? Our best bet is to try to detect one indirectly through its effects on other objects. A black hole has an exceedingly strong gravitational field. If it were created in the collapse of, say, a 20-solar-mass star, it would have the same gravitational field the star had (assuming it lost no mass as it collapsed). If this black hole was moving through a dense cloud of gas, it would pull in gas around it and we might be able to detect the resulting whirlpool.

Better still would be a binary, or double star system in which one of the components was a black hole. But if black holes are invisible, how would we know one was there? Astronomers are familiar with systems of this type; they are called spectroscopic binaries. They see only one star, but know that there are actually two, one in orbit around the another. To understand how they know this, let's begin with a brief discussion of spectroscopy. When the light from a star is passed through an instrument called a spectroscope, we get a series of lines called spectral lines. They give us considerable information about the star; in particular, they tell us how fast the star is moving toward or away from us. If the lines are shifted from their normal position toward the red end of the spectrum, we know that the star is moving away from us; this is called a redshift. If the shift is toward the blue end, the star is moving toward us and the shift is called a blueshift. This behavior is due to the Doppler effect. You are likely familiar with it in relation to sound. If a car approaches and passes you with its horn blaring, you hear a distinct change in the pitch, or frequency, of the sound. In the same way, the frequency of light changes when its source moves relative to you, and, as a result, the positions of the spectral lines change.

In binary star systems, the lines shift back and forth periodically. Actually, we see two sets of spectral lines, one from each star, superimposed on a single plate. Furthermore, if we took several photographs over a period of time, we would see that as one set of lines moved to the right, the other set would move to the left. This is what you would expect from two stars orbiting one another if we were seeing the orbit edge-on or at a small angle. In this case one of the stars would be approaching, the other receding.

But what would the spectrum look like if one of the two objects was a black hole? Rather than two sets of spectral lines we would see only one, but it would still move back and forth.

Once this was determined, astronomers scanned the skies looking for systems with spectral lines that behaved in this way. In particular, they wanted to find systems where one of the objects was invisible, yet massive enough to be a black hole. But they had little luck. Then, in December 1970, the first satellite devoted entirely to X-ray astronomy was launched off the coast of Kenya in Africa. It was christened UHURU (which means "freedom" in Swahali) in honor of Kenya's independence day. Within a short time several X-ray sources were found and by 1974 over 160 sources had been examined in detail; several were associated with binary systems. One that attracted considerable attention was in the constellation Cygnus; it was called Cyg X-1.

CYG X-1

Why would a binary system be a source of X rays? If one of the components of the system was a black hole, it would, under certain circumstances, seriously affect the other component. If, for example, it were pulling matter from its companion, this matter would swirl around the black hole in a disk, and as it plunged toward the black hole, considerable heat would be generated. The easiest way to understand why this happens is to consider an analogy with the solar system. In the solar system we know that the inner planets travel faster in their orbits than the outer planets. Earth, for example, travels faster than Mars. The same thing would occur in the swirling disk of matter around the black hole; the inner parts would move faster than the outer parts. But, because the disk is continuous, considerable friction would be generated as inner regions moved past outer ones. This would cause the matter to get hotter and hotter as it spiraled inward, until finally, just before it reached the surface of the black hole, it would be so hot it would produce large quantities of X rays.

Once Cyg X-1 had been identified as a particularly interesting source, astronomers began their search for the object that was creating it. The blue giant known as HDE 226868 was soon identified as the star closest to it, and within a short time Paul Murdin and Louise Webster of the Royal Greenwich Observatory in England obtained a spectrum of it. It was about 8,000 light-years from Earth and appeared to be approximately 23 times as massive as our sun. For the next several nights Murdin and Webster continued to take spectra. Soon it was evident that the lines were moving back and forth. HDE 226868 had an invisible companion and, within a short time, there was also evidence that the X rays were not coming from the star, but from its companion. Using an assumed mass for the blue giant, called the primary, they were able to show that the mass of the invisible component, called the secondary, was approximately ten times that of our sun. They were pleased: to be a black hole it only had to be three times as massive.

But there were problems. The mass of the invisible component was dependent on the mass of the primary, and although the spectrum of the primary indicated that it was a massive blue giant, it was quite possible that considerable mass had been eaten away by the black hole over the lifetime of the system. Furthermore, the mass of the secondary depended on the angle that we were seeing the system, in other words, the tilt of the orbit. We were obviously not seeing it edge-on; an eclipse of the X rays would occur if this was the case, and there was no evidence for an eclipse.

But at what angle were we seeing it? Astronomers are still not certain. To be on the safe side they knew they should use the case that gave the lowest possible mass. This is edge-on, but they knew we weren't seeing it edge-on, because there was no eclipse, so they assumed it was inclined at a small angle to this direction.

The Cyg X-1 system. Matter is being drawn off the primary (right) into an accretion disk (shown at the left). The matter in the accretion disk whirls into the black hole. Heating occurs near the black hole and X rays are emitted.

What about the X rays? What do they tell us? They do, indeed, give us an important clue. Rapid "flickering" of the X rays down to time scales of milliseconds (1/1000 second) have been detected. They tell us that the object is extremely small. To see why, consider a variable star, a star that varies periodically in light intensity. If the brightness of this star is to change, one side of it must communicate with the other to notify it of the change. The uppermost speed at which this communication can take place is the speed of light. This means that if the star is, say, one light-minute across, it can't vary in light intensity in less than a minute.

But Cyg X-1 is varying considerably in $1/1000$ second and can therefore be no more than about 100 miles across.

The swirling disk of matter that surrounds it is, of course, larger than this. It has been calculated to be several million miles across. Where, you might ask, does this disk come from? To answer this, it is best to begin by considering the Earth and the moon. Think of the Earth as the primary and the moon as the secondary. Both have gravitational fields; the Earth's is much stronger because it is more massive. But as you move upward above its surface, its gravitational field gets weaker. Nevertheless, for a given height, the field is the same everywhere. This means that you could draw a sphere around the Earth, say one mile above it, where the gravitational field is the same at all points. A similar sphere could be drawn at two miles and so on. These spheres are called equipotentials.

Now suppose that we continue outward in the direction of the moon. The Earth's field will continue to weaken as we move outward and eventually we will come under the influence of the moon; we will then fall toward the moon. Just before we reach this point, however, there will be a point where the two forces balance. If you now draw equipotentials around both the Earth and moon that pass through this point, you get a figure eight. One of the loops (the moon's) will, of course, be much smaller than the other (see figure, p. 171). This is called the Roche lobe and the point where the two loops join (where the gravitational field of the Earth and the moon are equal) is called the Lagrangian point.

Anything from the Earth that passes the Lagrangian point will be pulled to the moon. Similarly, anything from the moon passing in the opposite direction will fall to the Earth. HDE 226868 and its compact companion also have a Roche lobe and a Lagrangian point. And anything from HDE 226868 that passes this point falls toward its companion.

But how would anything from the primary pass it? There are two possibilities. When a star gets old, it runs out of fuel and begins to expand; in time it becomes a red giant. In another three or four billion years, our sun, for example, will expand out

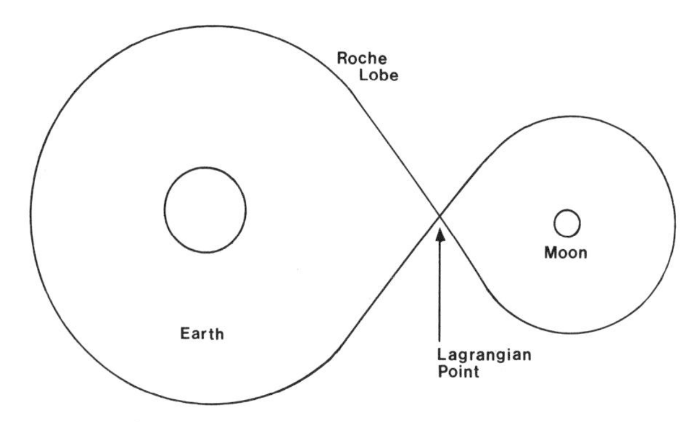

The gravitational equipotential called the Roche lobe of the Earth and the moon. The Lagrangian point is the point where the gravitational pulls of the Earth and moon are equal.

roughly to the orbit of Earth. If the star in a binary system such as Cyg X-1 expanded so its outer shell passed the Lagrangian point, material from it would spiral into its companion.

We know, however, that HDE 226868 is a blue giant, so it's not an old star. This brings us to our second possibility: large stars such as HDE 226868 have strong stellar winds, and much of the material that makes up this wind likely passes the Lagrangian point. It will not, however, drop directly toward the black hole. HDE 226868 is much larger than its collapsed companion; furthermore, it is spinning and, consequently, material from it will go into orbit around its companion. Gravitational and other forces will then shape it into an accretion disk. New material will continue to enter at the outer edge, but, because of sliding friction against inner layers, it will lose energy and spiral inward.

Most astronomers are now convinced that Cyg X-1 is, indeed, a black hole. It seems to have all the qualifications: it is invisible, emits X rays that indicate it is small, and it has an apparent mass greater than 3 solar masses. Recent work on it by D. R. Gies and C. T. Bolton of the University of Toronto slightly increases the earlier estimate of its mass.

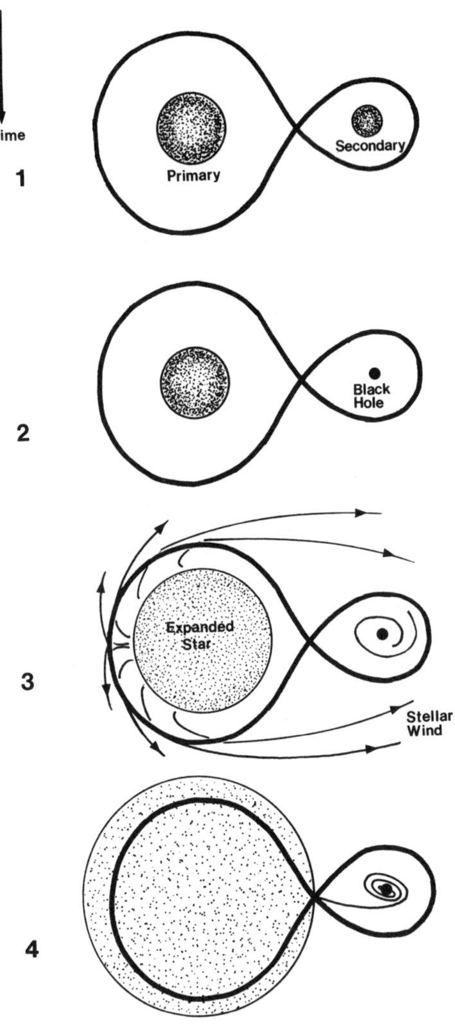

Evolution of Cyg X-1 to a black hole. A giant star collapses to a black hole (1,2). Stellar wind is dragged into the black hole (3). Eventually the star expands and is dragged into the black hole (4).

Bolton discussed their work with me. "Basically, what we did was obtain an improved velocity curve of the companion star," he said. "I think we've really tied it down extremely well, and we've used other information to establish its rotational velocity. A combination of these things raises its mass to between 12 and 20 solar masses."

I asked him what was the lowest possible mass it could have. "Others have shown it is about 6 solar masses. I think the things we have done, even though there are some assumptions involved, show that the very lowest number we could accept is about 12 solar masses . . . and 15 is more probable."

Cyg X-1 is still considered by most astronomers to be the best candidate we have today. But there are others, some almost as good.

LMC X-3

What is usually considered to be our second best candidate is an X-ray source in the Large Magellanic Cloud, an irregular galaxy seen only from southern latitudes. Called LMC X-3, it was shown to be a good candidate by Anne Cowley of Arizona State University and colleagues D. Crampton, J. B. Hutchings, and J. E. Penfold of the Dominion Astrophysical Observatory in British Columbia, Canada, and R. A. Remillard of MIT.

Cowley attended Wellesley College in Massachusetts, then went to the University of Michigan to study astronomy. I asked her how she first got interested in astronomy. "I took it in college, and liked it," she said. "But at that age you go by gut feeling, not by plan. At least I didn't go by plan. It just happened. At that time women didn't go to grad school, particularly in science." She hestitated for a few moments. "But I went . . . not sure it was going to work out. And, in fact, it worked out fine. After I graduated I worked at Yerkes Observatory for a number of years, then I went to Michigan, and now I'm here at Arizona State."

In the late 1970s, Cowley and her colleagues began a survey of

Anne Cowley.

the X-ray sources in the Magellanic Clouds, trying to identify and study as many of them as possible. Over a period of a couple of years they obtained the spectra of several possible candidates, but at this stage there was still considerable uncertainty. Then, in 1979, the Einstein X-ray satellite was launched into space and they were able to obtain much better positions for the sources.

Shortly after they obtained these new data, Cowley began going through some of their results for a paper they were writing. Looking over their data she noticed that the spectrum of one of the sources was different on two different plates. "I thought that was odd," she said. "I was sure it had to be wrong, so I started going through the literature and found that a Dutch group had also obtained a spectrum of the same source. And it was different from either of ours."

Cowley and Dave Crampton had time scheduled at the Cerro Tololo Inter-American Observatory in Chile, so they decided to follow up on the mystery by retaking the spectrum. In the fall of 1982 they traveled to Chile. "It was quite interesting the way things unfolded," said Cowley. "On the first night we took a series of spectra, and while Dave continued working with the telescope, I rushed down to develop the plates so we could have a quick look at what was going on."

Holding the developed plates up to the light she was amazed to find that there was a definite shift in the spectral lines. Returning to the telescope she and Crampton took more plates. Soon there was no doubt; the lines were not only shifting, they were moving back and forth periodically. The next night they took more plates, and got the same result. "By the second night we had measured the plates using a measuring device in the observatory and were sure that the variation was periodic . . . and the range was large. I could hardly believe it," she said. The velocity of the visible star was an incredible 146 miles/sec and it orbited the collapsed object in only 1.7 days. This meant that the invisible object had to be exceedingly massive, easily massive enough to be a black hole.

To be sure, they made further observations the following night. There now seemed to be little doubt, but they knew that more careful measurements would have to be made before they could announce their discovery.

"At supper we were sitting with a number of other astronomers, and everyone was chatting about what they were doing," said Cowley. "One of them turned to us and asked, 'What are you working on?' 'Oh, some X-ray sources,' I said. 'Have you found anything interesting?' he asked. It was quite funny . . . we were really hesitant to say we had found anything because you know how it is. You go home and reduce your data properly and everything goes away." She laughed.

They were sure the mass was greater than 3 solar masses; still, it was a relief when they took the data back to Victoria and checked them and the results were the same. "It was an exciting time," said Cowley, "because it's rare, at least in my experience, that as you go along you know exactly what you're getting. You may be getting good data, but you usually don't know what you're finally going to get. This was one of the few cases where we were pretty sure as we went along what we were getting. Everything just fell together. Things usually don't work out that well. People have said to me, 'you were sure lucky.' But the truth is I've got a drawer full of stuff that didn't work out." She chuckled, "I suppose everybody has."

Cowley and her colleagues have determined that the mass of the invisible companion is 9 solar masses. Since their initial work, they and a number of other groups have studied the source using photometric methods (light measuring). These measurements showed that the star is considerably distorted. Furthermore, they also allowed a mass determination and again a mass of approximately 9 solar masses was obtained.

Another important discovery in relation to this source is that the X rays from the source have a period of about 100 days. This appears to indicate that the accretion disk around the black hole is precessing in the same way a spinning top precesses; in other words, its spin axis traces out a cone.

I asked Cowley if it was difficult to distinguish LMC X-3's X rays from the other X ray sources in the Large Magellanic Cloud. She said that it is quite isolated, so there was no problem. "It's so far from the center of the Clouds, in fact, that one night the night assistant walked in while we were working on it and said, 'You're not even looking at the Large Magellanic Cloud.' "

Even though there is considerable evidence that there is ablack hole in LMC X-3, there are criticisms. It has been pointed out, for example, that the accretion disk around the X-ray source may contribute a significant portion of the optical emission from the system; if so, the companion would have a mass much less than 9 solar masses—in the extreme case, as low as 2.5 solar masses. Cowley said, however, that she isn't worried about this. "As far as I'm concerned the minimum mass you get from the velocities is considerably larger than the maximum neutron star mass, so it has to be a black hole," she said.

Cowley and her colleagues have also shown that LMC X-1, another X-ray source in the Large Magellanic Cloud, may also contain a black hole. She admits, though, that there is considerable uncertainty in this case. "LMC X-1 is in a complicated region of the Cloud," she said. "It's right near a big complex that has a lot of X-ray emission, and it's also close to the 1987 supernova, so it's in a crowded field. It's not as straightforward as LMC X-3." Using the data they have, they have determined the mass of LMC X-1 to be approximately 4 solar masses.

Recently the group has discovered an even more interesting candidate in the Large Magellanic Cloud. It is unlike any other candidate in that it eclipses its X rays, and is therefore seen edge-

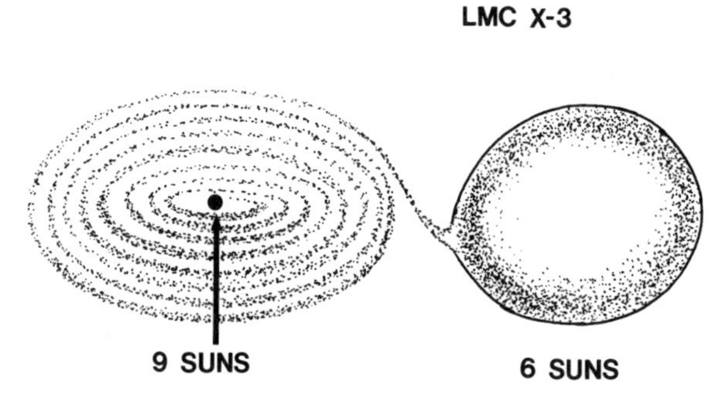

The LMC X-3 system according to best estimates. The black hole has a mass of 9 solar masses (suns). The primary has a mass of 6 solar masses. The left-hand side shows the accretion disk around the black hole.

on. Called CAL 87 (Columbia Astrophysical Lab source 87), it's a binary X-ray source with a period of ten hours. "It clearly has a very massive secondary," said Cowley. "What you see is a combination of a bright accretion disk around a collapsed object and a star something like our sun. The unseen object has a mass of about 6 to 7 solar masses, so it's very massive. The fascinating thing about it is that it will allow us to determine what the structure of an accretion disk near a black hole is really like because we are seeing it edge-on."

A0620-00

Another candidate, known as A0620-00, was discovered by Jeffrey McClintock of the Harvard–Smithsonian Center for Astrophysics and R. A. Remillard in the mid 1980s. The discovery was a result of their search for a binary system where the invisible object was much more massive than the visible one.

McClintock did his undergraduate work at Stanford in the early 1960s, then went on to MIT for a Ph.D. His interest in astronomy began in junior high school when he built a telescope and began observing the stars. He said that he forgot about astronomy after he went to university to study physics, but in graduate school he became interested again when he began doing X-ray astronomy. His thesis was on the X rays in the giant elliptical galaxy M 87.

McClintock pointed out the advantages of a system like A0620-00 over one like Cyg X-1. "A system like Cyg X-1 is not what you really want," he said, "because the massive star doesn't orbit the black hole; the two objects orbit a point roughly midway between them—their center of mass. So by just measuring the velocity of the supergiant and determining its orbital period you don't prove it is a black hole. You have to go further. You have to argue about how massive the visible star is." He paused. "Our

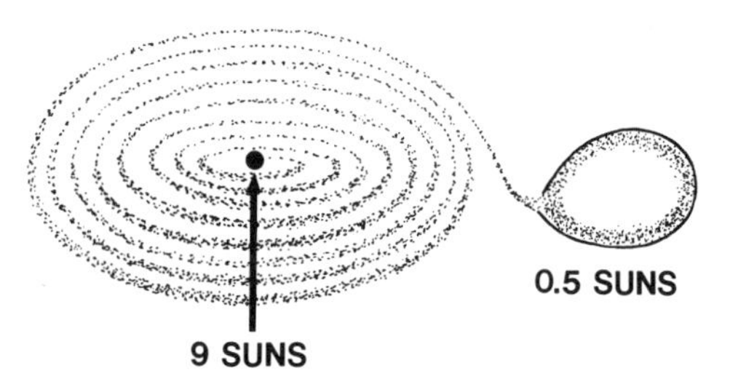

The system A0620-00. The primary is a small orange star. The black hole has a mass of 9 solar masses.

idea was: if we could measure the radial velocity [velocity component in our direction] of a very-low-mass star, orbiting an object that had, say, 10 solar masses, it would be like having a little satellite going around the Earth. A little star going around a compact object would give you a direct measure of its mass [and tell you if it was a black hole]."

There is, however, a problem with the most promising systems of this type. They are very bright X-ray sources, and the X rays produce a lot of heat. This heated gas gets in the way, and you therefore can't see the absorption lines (dark spectral lines) from the star clearly. This means you can't determine its velocity accurately. McClintock and Remillard looked for a way to get around this problem.

In 1980 they began going through lists of X-ray sources, searching for one that might have a small-mass primary. A few years earlier, in 1975, A0620-00 had been one of the brightest X-ray sources in the sky. As a bright source it would have been of little interest to them, but over a period of about two years, its X-ray emission had stopped. It soon came to McClintock's attention. "It was worth serious consideration from three points of view," said

McClintock. "First, it had a low-mass companion. Second, it was very close to the Earth, only about 1 kiloparsec [3200 light-years] away. And third, it had been a brilliant X-ray source so you knew it had a massive compact object, but the X rays had turned off."

McClintock and Remillard went to work on the system. Within a few years they found that the companion, a dim orange star with a mass about one half that of our sun, had an extremely high velocity around the visible object, almost one million miles per hour (284 miles/sec). At this speed it completed an orbit in one third of a day. By comparison the Earth travels at only 18.5 miles/sec in its orbit and takes 365 days to complete an orbit.

"What is particularly nice about A0620-00 is that with just the orbital period and the velocity of the visible companion you get 3 solar masses for the worst possible case—when the orbit is edge-on," said McClintock. But we know it's not edge-on since there were no eclipses when it was an X-ray source. Some evidence points to a tilt of about 45 degrees to our direction. If this is the case, the collapsed object has a mass nine times that of our sun, and is, indeed, a black hole. On the other hand, if it is tilted, say, 15 degrees more than that, it would have a mass of 16 solar masses. From this point of view it certainly looks like an excellent candidate.

"I'd sure like to find another object like A0620-00 that has an even higher velocity secondary, with a short period," said McClintock. "If you could find a system for which the two numbers, orbital period and velocity, gave, say, 6 solar masses, you would know right away it was a black hole. There would be no doubt [with just these two numbers A0620-00 gives 3 solar masses, too close as far as McClintock is concerned]. That would be a major improvement . . . but there are very few systems of that type."

McClintock and Remillard are still working on A0620-00, making refinements and so on. In their first paper they had only ten velocity measurements; in a recent paper they published they had 44 independent measurements.

SS 433

Although it is not as strong a candidate as the above three, SS 433 is a particularly intriguing system. It is the 433rd object in a catalogue of stars with emission (bright) lines compiled by C. Bruce Stephenson and Nicholas Sanduleuk of Case Western University in the mid 1960s. Stars with bright line, or emission, spectra are interesting because they are relatively rare. Most stars show only dark lines.

No one paid much attention to SS 433 at first. It was only after a series of seemingly unrelated events that it was noticed. First, in 1976, F. D. Steward and several associates noticed a weak X-ray source in the data of the British X-ray satellite, Ariel 5; it was in the constellation Aquila. A year later David Clark, A. J. Green, and J. L. Caswell noticed a variable radio source in Aquila. Shortly thereafter a supernova remnant called W50 was found at about the same position.

All of these events were unconnected until David Clark and Paul Murdin discovered that SS 433 and the radio source in Aquila were one and the same. They wondered if it was connected with the supernova remnant W50. If so, it would be a particularly interesting object. They took its spectrum, but didn't find anything out of the ordinary.

Bruce Margon of UCLA then entered the picture. Margon and several associates began taking spectra of the object in August 1978. Soon it became obvious that this was no ordinary object after all. The expected emission lines were there, but several unidentified weaker lines were also visible. The group continued taking spectra, and within a few days it was obvious that the weak lines were moving. In fact, they were moving so rapidly that you could see the shift over a period of a single night. Margon was astounded. It was a huge shift, indicating a velocity of 50,000 kilometers/sec (about 30,000 miles/sec). But of more interest was the fact that all of the lines were not moving in the same direction: some were moving toward one end of the spectrum and others were moving toward the other end. The object was both moving toward and away from us. It seemed impossible; speeds of 30,000 miles/sec had been seen before, but only in distant galaxies and quasars. And there was no possibility that SS 433 was either of these: it was only 11,000 light-years away.

When more complete data were obtained, it was found that, in all, there were four sets of spectral lines, all moving differently. Nothing like this had ever been seen before. There was no doubt: it was an incredible object. But what was it?

Margon announced his results in December 1978 and they created a sensation. What could possibly cause such strange behavior? One of the first suggestions came from Andrew Fabian and Martin Rees of Cambridge. They suggested that the objects had two jets, one of which was pointing in our direction and the other away from us.

Many people jumped on the bandwagon, but observations had just begun when it disappeared into the day sky. In the spring of 1979, however, it was back again in the night sky and Margon and his team went to work on it immediately, as did several other groups. Within a short time, a 164-day period was discovered for one of the sets of lines.

A number of astronomers offered alternate models. But calculations showed that most of them were impossible; a black hole with millions of solar masses would be needed. The only thing that seemed certain was that it was a binary system with an accretion disk around one of the two objects.

The model that is now generally accepted is due to Bruce Margon and George Abel of UCLA. It is usually referred to as the "kinematic model." In it the accretion disk is ejecting two oppositely directed beams of matter along an axis inclined to a second axis of precession (see figure on next page). The beam precesses around this axis with a period of 164 days. There is also a slight eclipse of the accretion disk.

But what is the mass of the invisible component? Is it massive enough to be a black hole? As in the case of Cyg X-1, its mass depends on the mass of the visible star, which appeared to be

SS 433. Matter is being dragged into a black hole from a large nearby star. It is being ejected out in two jets.

about 20 solar masses; if so, the mass of the invisible component is 7 solar masses, easily massive enough to be a black hole. But as in the case of Cyg X-1 and others, problems persist. For example, it has been suggested that the compact object could itself be a close binary system. If this is the case, both could be neutron stars. But few believe it is; most of the evidence at the present time seems to indicate a black hole.

An important question that we haven't answered so far is: What causes the jets? One suggestion is that matter is coming into the black hole so fast that it can't absorb it, and as a result it squirts

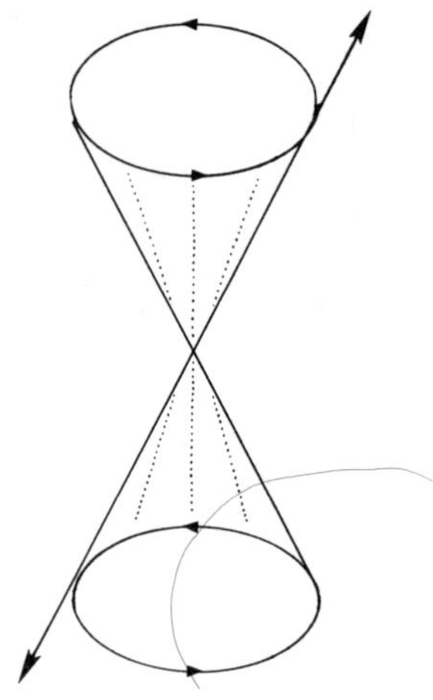

A simple representation of the precession of the jets of SS 433. They trace out cones.

it out. This may, in fact, be an early phase of all systems of this type. Cyg X-1 and similar systems may go through such a phase, and later, as the black hole is able to absorb the material, the jets disappear.

GALACTIC BLACK HOLES

While most of the attention in the last few years has centered on stellar mass black hole candidates, many astronomers are now convinced that supermassive black holes—millions and even billions of times as massive as our sun—may reside in the cores of galaxies. Radio galaxies and quasars generate tremendous energies, and supermassive black holes seem to be the only way of explaining it.

Galaxies such as M87, Cygnus A, and Hercules A and quasars such as 3C-272, all have jets emanating from them. In some cases these jets extend out hundreds of thousands of light-years. Something in the core of the galaxy must be causing them. The current best model of this activity assumes a massive black hole at the core surrounded by a gigantic accretion disk. Material near the black hole is being ejected outward in opposite directions through a central hole in the accretion disk.

One of the first to suggest such a model was the British astrophysicist Donald Lynden Bell (1969). In his original model, he assumed the black hole was nonrotating. But, in 1970, James Bardeen of Yale University showed that black holes in galaxies would probably rotate, and interest soon centered around spinning, or Kerr, black holes.

A case of particular interest was the elliptical galaxy in Virgo, called M87. It has a jet about 6,500 light-years long that is a strong source of both radio waves and X rays. In 1979, Peter Young and Wallace Sargent of Caltech took a close look at M87. They measured the light intensity across the jet and took spectroscopic measurements. The center was exceedingly bright, indicating a high density of stars, and spectra showed that their velocities were incredibly high. The only way to explain this was by assuming that a massive black hole—up to 5 billion solar masses—was at the core.

A galaxy is active, however, only if the black hole has a source of fuel, namely stars and gas. If none are available, there would be little evidence for a black hole, and this is likely the case in older sources where the fuel has been depleted. This means that ordinary galaxies, such as those around us, may also have a black hole at their core. And, indeed, we have evidence that this is the case. As early as 1971, Martin Schwarzschild of Princeton University showed that the stars near the core of the Andromeda galaxy whirl around in a flattened disk. More recently, Alan Dressler of the Carnegie Institute and several colleagues have shown that the velocity of these stars increases rapidly as you approach the center. Using these data they were able to show that a mass of 30 to 70 million solar masses would be needed to make the stars go this fast.

Then, in 1984, John Tonry of MIT showed that there is a

The elliptical galaxy shown may have a gigantic black hole at its center. (Courtesy National Optical Astronomy Observatories.)

similar increase in the core of the small companion of the Andromeda galaxy, called M32. According to his calculations, it likely harbors a mass of about 5 million solar masses in its core. Furthermore, the nearby galaxy known as the Sombrero galaxy also has a dramatic increase in stellar velocities near its core, indicating a mass of about a billion solar masses.

But are these masses necessarily gigantic black holes? McClintock believes they are. "It's much harder to talk about the center of a galaxy than it is stars that are near us," he said. "But it certainly seems that something strange is going on. To whip the matter around as fast as it appears to be going you definitely have to have a large mass. But it's a very complex problem. If you want to establish the case for galaxies, you first have to get your hands on nearby black holes, such as those in X-ray systems, and nail them down. Prove they are black holes first . . . then go to galaxies and study them."

Anne Cowley replied to the same question with, "I think it's hard to understand the velocities without having massive black holes. My personal feeling is that they are not as well established as black holes in binary systems because you don't have dynamical information. It's hard to argue against Kepler's laws. The arguments for black holes in the centers of galaxies are more indirect, but it's probably the only interpretation unless there's some really funny physics going on that we don't understand. I guess that's a possibility, but I don't think it's likely."

And finally, Tom Bolton of the Universty of Toronto: "I think there's a plausible case for it. For the most part, though, it's weaker than it is for the stellar case. But it is still quite likely that there are massive black holes there. However, we need more data to prove it."

IS THERE A BLACK HOLE IN THE CORE OF OUR GALAXY?

As I mentioned earlier, a black hole by itself does not give off energy; it needs a supply of fuel, and this likely comes from the stars and gas that are whirling around it. But eventually these stars and gas are depleted and the black hole runs out of fuel. Many astronomers believe that is the situation in our galaxy.

Compared to galaxies such as M87 and Cygnus A, our galaxy is pretty bland. But we are relatively close to its core and to us it appears to be relatively active. Ordinary telescopes cannot penetrate the gas and dust between us and the core and consequently are of little use to us. Infrared telescopes, on the other hand, are able to penetrate them; as a result, we now have a fairly good idea what this region looks like.

Our galaxy, like other spirals, has a large nuclear bulge. Near the center, extending about 10,000 light-years out from it is a turbulent disk of hydrogen that rotates at high speed. Inside this disk are molecular clouds and, finally, at the center is a small compact source called Sgr A* that gives off considerable radiation.

Looking at the core of Sgr A* we find a ring of whirling matter about 12 light-years out from its center. Temperatures are high in this region, 10,000 to 20,000 degrees. Furthermore, the density of stars is high here and they travel at high speeds.

Is there a black hole here? There are strong indications that there is and many astronomers are convinced of it, but so far they don't have proof.

WHY ARE THERE SO FEW BLACK HOLE CANDIDATES?

One of the things that may strike you when you look at the black hole candidates we have is: Why are there so few? We have several hundred good neutron star candidates and black holes are created in the collapse of a star only slightly more massive than those that create neutron stars. In fact, the two objects aren't that different in size.

McClintock believes the reason is the "no hair" theorem we talked about earlier. It tells us that black holes have few properties, only mass, spin, and electric field. "That's the problem in a nutshell," said McClintock. "With neutron stars you can have magnetic field lines giving strong fields of about 10^{12} gauss, with charged particles entering them. Matter flowing down onto the rotating neutron star from another star gets whipped around like an eggbeater. Furthermore, a neutron star has a hard surface. You can go up and give it good knock. There's nothing like that in the case of black holes. There are a few hundred neutron star binary systems, and about 50 of them are known because the X-ray emission from the system is pulsed every second, or ten seconds, or whatever. And that's an . . . [indisputable] signature. You know you've got a neutron star right away. Nothing else is going to do that. Black holes can't do it. Also, objects that aren't even X-ray binaries, just lone objects like radio pulsars, sit out there and make beautiful radio pulses. If you had a solo black hole out there, it couldn't do that."

Bolton is convinced that it's a combination of circumstances. "First of all, black holes are probably only formed from the most massive stars," he said, "and there may be situations where the most massive stars manage to avoid becoming black holes. Massive stars are rare, that's the first problem. The second may be that there is only a short period of time for any binary system where the black hole will call itself to our attention by emitting X rays. So I think the probability of having a black hole in our vicinity, close enough to study, is pretty low."

Despite the relatively small number of candidates, it does seem that we have several good candidates and that black holes actually do exist. If so, our next problem is: Can we use black holes as tunnels to distant points of the universe? What, in fact, would it be like to go into a black hole?

CHAPTER 10

Journey into a Black Hole

If black holes do exist, it is quite possible that we will visit one some day in the distant future. Even now we are capable of sending rockets to the most distant planets in the solar system. And recently a rocket exited the solar system and is now headed for the stars. It will, of course, be hundreds of thousands of years before it reaches the nearest stars; nevertheless, it's a start.

Mankind has obviously accomplished a lot in a relatively short period of time and it seems likely that our technology will continue to advance. With this in mind, it's interesting to contemplate what we will be capable of 1,000 or even 10,000 years from now. It is certainly likely that we will have the technology to visit nearby stars and black holes.

Regardless of how advanced our technology becomes, however, we still have a serious problem to overcome. According to the theory of relativity we cannot travel faster than the speed of light. Yet most stars of interest are tens, hundreds, and even thousands of light-years away. This means that even if we could travel close to the speed of light, it would take many generations for a trip to one of them. Somehow we have to get around this problem and the best prospect for doing this at the present time is through space-time tunnels. Let's assume, then, that we could travel into one of these black hole tunnels. What would it be like?

JOURNEY INTO A BLACK HOLE

What would we see if we went into a black hole? The best way to find out would be to use a probe. We could, for example, go into orbit around a black hole at a safe distance, then send a small probe into it.

What is considered a safe distance? You would obviously have to be far enough away so that the tidal forces were weak. We know that on Earth we do not feel any tidal forces from the sun. In fact, if the sun suddenly collapsed and became a black hole, the gravitational pull on the Earth would not change. The Earth would just continue going around it in the usual way. But the Earth is 93 million miles from the sun; we would certainly want to get much closer to the black hole than this.

How close could we get without being pulled apart? That depends on the mass of the black hole. Let's assume it is approximately ten times as massive as the sun. A simple calculation shows that in this case we could approach as close as 2,500 miles without experiencing any disastrous effects, but we would have to be very careful. The gravitational field of the black hole at this distance would be extremely strong. But even at this distance we would still need a telescope to see the black hole. It would barely be visible as a small dark region, seen only because it blocks off background stars.

From this orbit, however, we would be able to send a small probe to study it in detail, much in the same way we sent probes down to the surface of the moon during the early lunar flights. The probe could be equipped with TV cameras, clocks, radiation detectors, and so on.

As we sat safely inside our orbiting spacecraft we would be able to receive information from the probe as it spiraled into the black hole. And in the process we would find out what it would be like to go into a black hole.

We know that from the point of view of a distant observer, such as someone in an orbiting spacecraft, that a clock aboard the probe would appear to run slower and slower as it approached the

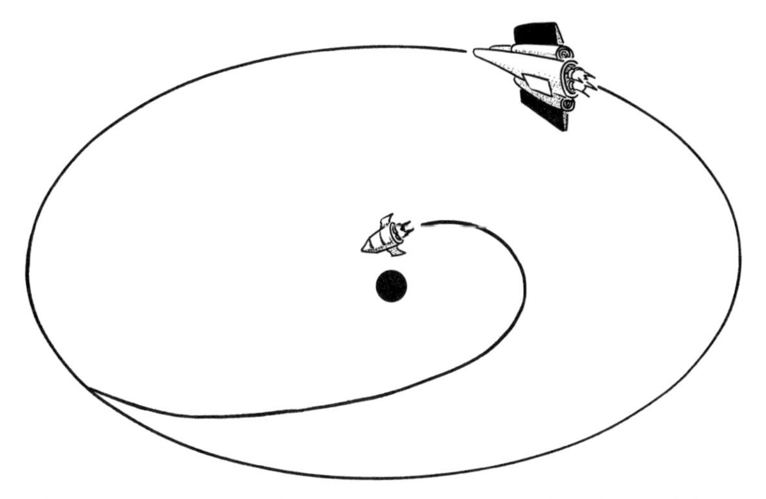

Rocketship orbiting black hole. A probe is sent down to enter and examine the black hole.

black hole. Furthermore, as it entered the increasingly strong field of the black hole, it would become more and more difficult for photons of light to escape. This also applies to the TV messages being sent back.

Photons must do considerable work to escape the high gravitational field, and therefore they lose a large amount of energy. It might seem that this would cause them to travel slower, but this isn't the case. Radiation of any type can only travel at one speed—the speed of light. What happens is that there is a shift in the frequency of the light to longer wavelengths. And this, in turn, causes the light to get redder.

In addition to receiving the signal from the probe, we could also observe its progress through a telescope. This would give us information about how well it was standing up to the tidal forces. But of most importance is that we would see what it would be like to enter a black hole.

To some extent we already know what this trip would be like. A trip into a black hole has been simulated on a computer. In 1975, C. T. Cunningham of Caltech used a computer to determine what it would be like to go into a Schwarzschild black hole. He was mainly interested in what the sky around the black hole would look like. In his calculation he assumed that a rocketship with an observer aboard free-fell into a black hole. At various points along the trajectory into it the observer presumably stopped his rocketship momentarily using retro-rockets and snapped photographs in the forward and backward directions.

Let's consider what you would see if you were this observer. Initially, the black hole looks like a small black dot, seen only because it blocks off light from background stars. At this stage, the sky is, for the most part, undistorted. As you approach closer, the black hole appears to increase in size and all the stars around you appear to move forward. As you get closer, the stars begin to form rings around the black hole. These rings are actually images of one another, each ring contains all of the stars seen around you when you are a long distance from the black hole.

As you approach the black hole closer, it fills more and more of the sky. Large numbers of bright rings are now seen around it. As you pass the photon sphere the black hole fills the entire sky in front of you. Moving even closer you find the darkness beginning to surround you. Looking back over your shoulder you see the series of bright rings. This is the last you see of our universe. Finally, as you approach and enter the event horizon, the bright rings merge into a point, then suddenly you are completely surrounded by darkness.

This is the story for a Schwarzschild black hole, but there are, of course, other types of black holes. William Metzenthen of Monash University in Australia looked into the case of the charged black hole in 1990. A charged black hole is different from a Schwarzschild one in that it has two event horizons.

The results from Metzenthen's computer calculations are shown in the figures. Outside the event horizons his results were similar to Cunningham's. As you approached the black hole a

A rocketship entering a black hole. Lines represent curvature of space (tunnel) leading up to the black hole (artistic conception).

A closer view of the black hole. Bright rings eventually begin to form around the black hole.

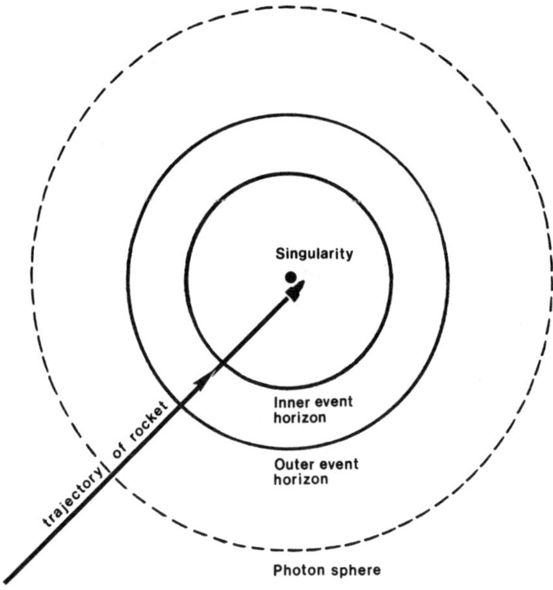

Trajectory of the rocket traveling into the black hole. Photon sphere and outer and inner event horizons are shown.

series of rings would form around it. Metzenthen found, however, that things were quite different inside the event horizons. He found that a series of "reflected" images would be seen inside the inner event horizon. A series of bright rings would therefore be visible. And as you continued closer to the singularity these rings would increase in size and merge as shown in the figures (pages 198–202).

As interesting as it is to imagine entering a black hole, we have to ask ourselves if it's really likely that we will ever be able to use these tunnels for space travel? What about the problems we talked about earlier? We saw, for example, that the tunnels are unstable and tend to "pinch off" before we can get through them. The "stretching" forces near them are also so great that you would

a

(Pages 198–202) A computer-generated sequence of views as seen by someone in a rocketship entering a charged black hole. The figures represent constellations in the sky. You can think of each of them as a series of stars. For simplicity they are represented in the form of cats and stars (remember, the constellations out there are different from those seen on Earth). Each figure shows the entire sky around the observer.[See figures (a)–(h).](a) The spaceship is a long way from the black hole (1,000 times the gravitational radius). The black hole is at the center of the diagram, but cannot be seen. The apparent distortion seen at the outer edges is due to the fact that the constellations are represented as if they were seen on the inside of a dark sphere. The outer ones are actually behind you. (b) Closer to the black hole. The black hole is directly in front of you. The dotted line is perpendicular to your view. (c) The black hole is now visible in the sky before you. A bright band of light is starting to form around it as constellations appear to move forward. (d) The spaceship is now passing through the photon sphere (see figure on page 200). (e) The spaceship has now passed through the outer and is just passing through the inner event horizon. The black region fills a large part of the sky before you, and there appear to be rings of light around it. They are actually the constellations we saw earlier. Each ring is a duplicate of the one inside it. (f) The spaceship is now inside the inner event horizon. The white spot at the center is a result of images from the outside that are reflected inside. (g) Very close to the singularity. Numerous rings of light appear before the spaceship. (h) A final view. External universe is now completely cut off. Images are considerably distorted.

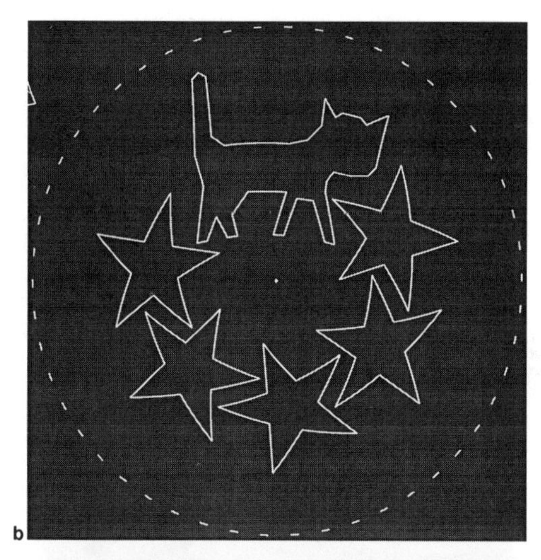

b

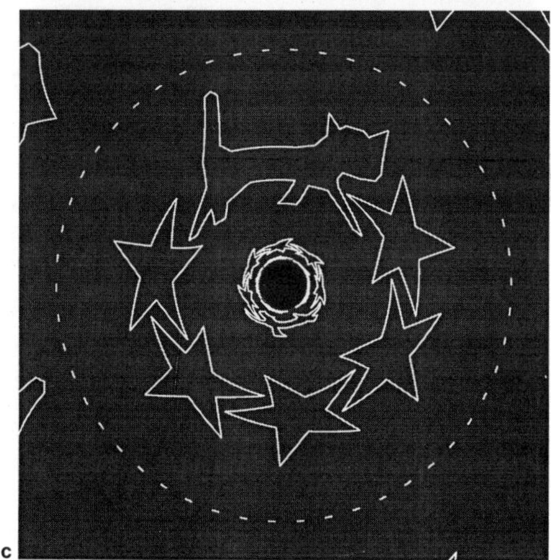

c

(continued)

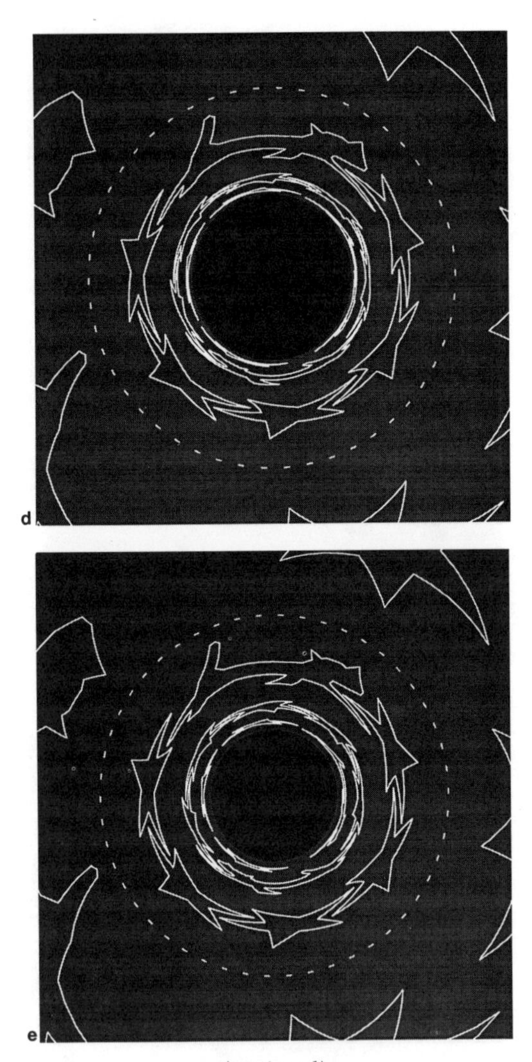

(continued)

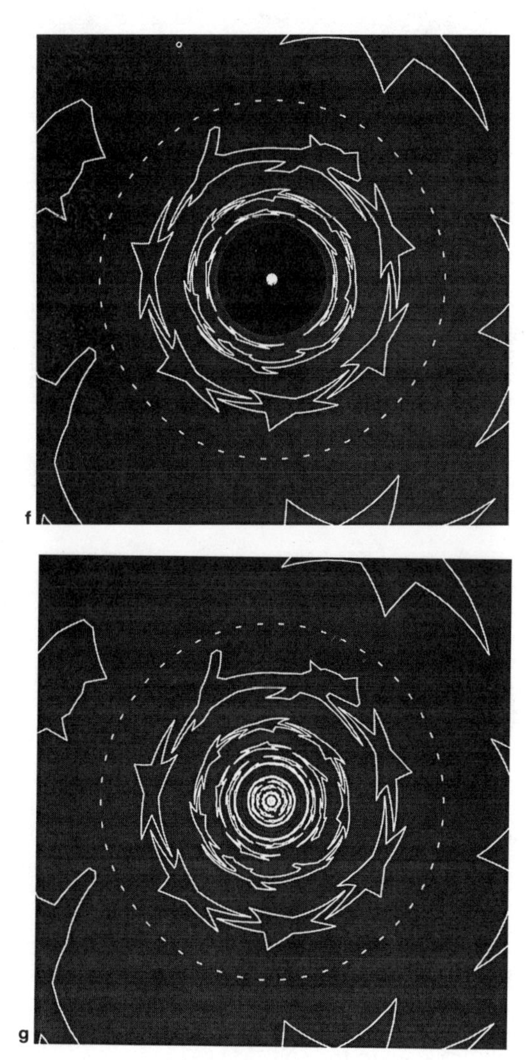

(continued)

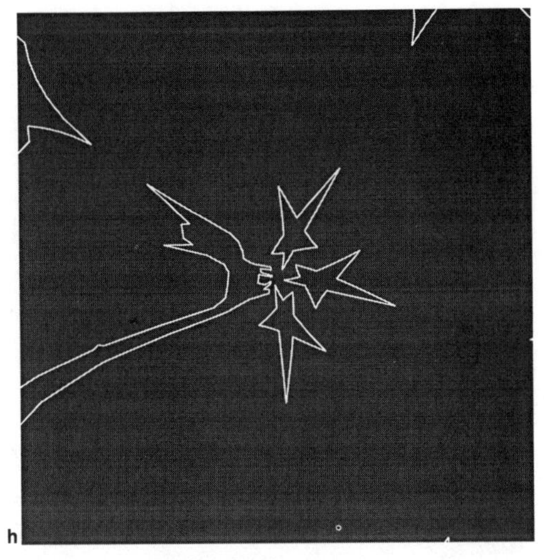

(continued)

be literally pulled apart before you got into one. Furthermore, there are problems with high levels of radiation and exits. And, finally, all black holes have event horizons so that the tunnels associated with them would be one-way. (You can only go in through an event horizon; you cannot pass out through one.) Can these problems be overcome? Needless to say, they are serious, so serious that we'll take a look at each of them in detail.

FUNDAMENTAL PROBLEMS

Black holes do, indeed, have tunnels associated with them. These tunnels are seen in the solution of Einstein's field equations. We saw earlier, though, that the tunnel associated with a Schwarzschild black hole is of little use to us. It takes a speed greater than that of light to get through it. Kerr black hole tunnels, on the other hand, are passable with a speed less than that of light. Even so, as we just saw, there are problems.

First of all, the prediction that there is a tunnel associated with these black holes comes from general relativity. Israel, Hawking, and others proved in the 1970s that a massive spinning star will eventually collapse and become a Kerr black hole. But their proof applied only to the region outside of the event horizon. General relativity tells us nothing about what goes on inside the event horizon. According to the equations, there is a tunnel leading up to the horizon, but what happens beyond that point is uncertain.

Does the tunnel extend past the event horizon? "It's unclear what the interior solution is," said Michael Morris of the University of Waterloo, who has been working on black holes for several years. "You can get a solution by extrapolating the space-time. . . . The Kerr tunnel looks like a Kerr black hole from the outside, and in the interior you can assume it is a tunnel. The problem, though, is that you're playing with the topology of space-time . . . you're not sure. You're on the outside of the black hole, and you're not really sure there is a tunnel, and if there is, whether it goes anywhere."

Morris pointed out that even if there were a tunnel, it would be very unstable. Literally anything that disturbed it would cause it to pinch off. A spaceship would not be able to pass through it. As it entered the tunnel it would create a disturbance, causing it to pinch off.

What causes this pinch-off? Basically, it's just something that comes out of the equations. There's no way we can really explain why it occurs. It's inherent in the makeup of the tunnels, and as such is obviously a serious problem. Furthermore, it's not the only one.

The tunnels associated with black holes have event horizons—surfaces that only allow you to pass through in one direction. Once you have passed through such a surface, there is no way you can get back again. In effect, it takes a speed greater than the speed of light to get back out, which we know is forbidden by relativity.

This means that if we were somehow able to pass through a black hole tunnel and ended up at some distant point in the

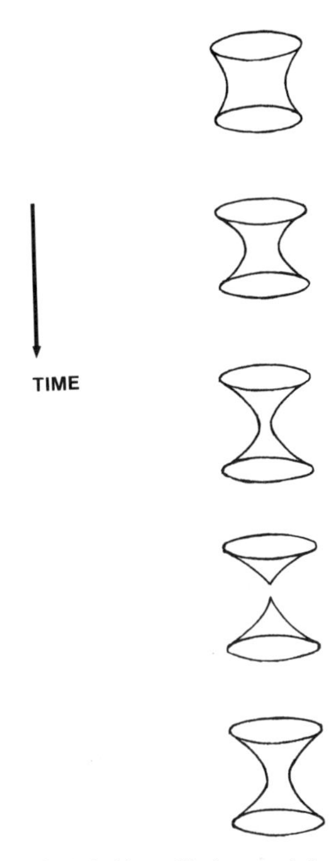

The "pinching off" of a wormhole.

universe, we would not be able to come back through the same tunnel. We would have to use a different one. But the probability that there would be a tunnel nearby that would bring us back to our starting point is obviously very small. Black holes, in general, are widely separated. If you think about it, in fact, there's another problem. If we passed through a black hole tunnel and exited, it's unlikely that we would know exactly where we were in the universe. So if we couldn't come back through the same tunnel, in all probability we would never get back to where we started.

Another problem with black holes involves transit times. As we saw in an earlier chapter, an external observer, say, an astronaut waiting in a spaceship, would see that the time it takes for another astronaut to fall into a black hole is infinite. He never appears to quite reach the surface. This means that a journey through a black hole tunnel, according to an external observer, would take forever. Of course, things are different for the astronaut who is actually going into the black hole. Only a small finite time elapses on his watch before he passes through the event horizon. And, according to his watch, he passes through the tunnel in a short period of time.

TIDAL FORCES

Another problem is the force you would feel on your body as you approached the black hole. To understand how it arises, let's begin with the Earth and moon. You are no doubt familiar with the fact that the moon causes tides on Earth. When the moon passes over an ocean, it pulls the water outward, causing it to bulge. Not only is water pulled toward the moon, but when the moon is over land, there is also a slight bulging of the land mass in its direction. In the same way, there is a bulge in the moon's mass in the direction of Earth. Furthermore, if the moon were closer, this bulge would be greater. Basically, it is due to the fact that the Earth is pulling on the side of the moon that is facing it with a much greater force than it is on the side away from it. (Gravity drops off with distance and therefore pulls more on nearby objects than it does on distant objects.)

Because of this differential pull, the moon is being pulled apart. If it were closer to the Earth, it would be pulled apart with an even greater force. This means that if you moved the moon closer and closer to the Earth, it would become increasingly oblong in shape, until finally the forces pulling it apart exceeded those keeping it together. It would then shatter into millions of tiny rocks, and in time these rocks would end up as a ring around the Earth's equator. Astronomers believe that this is how Saturn got its ring (it is also possible that some of the matter of the ring is material that was left over when Saturn was formed).

The forces that cause this breakup are called tidal forces. To some degree we experience them on Earth. The gravitational pull on your feet, for example, is greater than the pull on your head. The difference, however, is so small that it causes you no discomfort. If you were standing on the surface of a more dense object, however, it could be significant. Near the surface of a neutron star, for example, you would be pulled apart. And since a black hole is even more dense than a neutron star, you would not be able to approach it closely.

This means that we would experience tremendous tidal forces if we attempted to enter the tunnel associated with a black hole. In a stellar collapse black hole these forces would easily pull us and our spaceship apart. Is there any way around them? It turns out that there is. As the mass of the black hole increases, the tidal forces decrease; for a mass of 10,000 solar masses, in fact, the tidal forces at the horizon are negligible and you could easily pass through without any discomfort. But the gravitational radius of a 10,000-solar-mass black hole is over 5,000 miles. (The gravitational radius of a stellar collapse black hole is, by contrast, typically only a few miles.) This massive black hole would therefore have a tunnel that was tens of thousands of miles across. And of course this isn't what we normally think of when we think of a space-time subway to distant points of the universe. Furthermore, about the only place we would find a black hole with a mass of 10,000 solar masses is at the core of a galaxy. This means that our nearest black hole tunnel with moderate tidal forces would be in the core of our galaxy. And if we wanted to use other tunnels we would have to go to other galaxies, which would obviously defeat our purpose. We need space-time tunnels to travel long distances and these tunnels are separated by millions of light years. If this were the case, it wouldn't make much sense to try and use them.

Effects of tidal forces. The top astronaut is a long distance from the black hole. As he get closer (bottom), tidal forces stretch him until he finally resembles a piece of string.

EXITS AND WHITE HOLES

As if we don't have enough problems, there is another even more serious one. According to scientists, if we were to use one of these tunnels, we would enter the end associated with the black hole, pass through it, and exit at some distant point in the universe. But this brings up the problem of an exit. Black holes only pull things in; once inside the event horizon you can never get out. So, how do you exit? There is, indeed, a solution to Einstein's equations that shows that exits can exist. Scientists refer to the exit end of a black hole as a white hole. A white hole would, in effect, be a "gusher"; it would emit particles and matter, but nothing could pass in through it.

Do we have any observational evidence for white holes? Some of the objects in the outer reaches of the universe do appear to be ejecting material. Quasars and Seyfert galaxies both seem to be gushers. But if white holes do exist, we would have to ask how and where they were formed. The most likely place is, of course, the early universe. Shortly after the big bang explosion, the universe was extremely hot and in a state of rapid expansion. Some regions

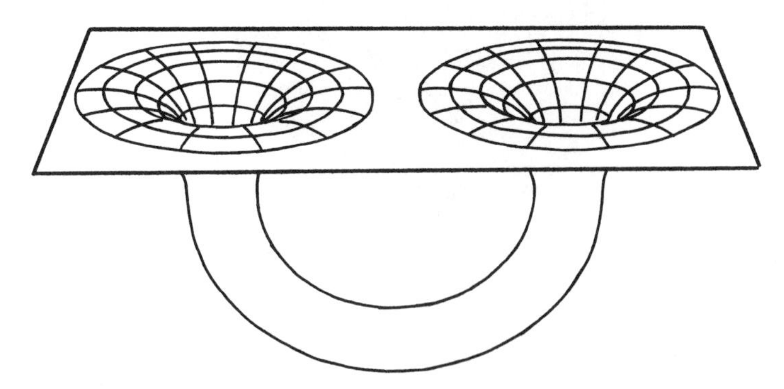

A wormhole in space. One of the ends is a black hole, the other a white hole.

were, no doubt, left behind in the early moments of this explosion. We refer to such regions as "lagging cores." These lagging cores would appear to us now as white holes.

Doug Eardley of Yale University has studied the formation of white holes in the early universe. Strangely, his calculations show that, even if there were lagging cores in the early universe, they would not become white holes. He demonstrated that they would attract a considerable amount of radiation—so much, in fact, that as radiation piled up around the core it would begin to warp the space around it. Within 1/1000 of a second this warping would become so severe that the lagging core would be transformed into a black hole. In short, any white hole that began to form in the early universe would be quickly converted into a black hole. This means that we would have entrances to our tunnels, but no exits.

RADIATION PROBLEMS

Let's assume that we are somehow able to overcome the above problems and travel into the tunnel. Would there be any further hazards awaiting us? Indeed, there would be. As we saw earlier, Hawking showed that black holes give off radiation and particles. Particle pairs are, in effect, literally ripped out of the vacuum near the event horizon. If you approached this region, these particles and radiation would be lethal. Fortunately, though, this is a serious problem only if the black hole is extremely small, and since we have to assume our black hole is very massive to overcome tidal forces, this radiation would not be a problem.

But our spaceship would have to pass through the event horizon of the black hole and into its interior. And at the center of the interior is a singularity. If it is a Kerr black hole, this singularity will be ring-shaped. Furthermore, the space near this singularity is even more warped than it is near the event horizon. This means that a considerable amount of radiation and particles will be produced near the singularity. According to quantum theory, large fluxes of particles and radiation will be spewn all around it with an intensity so great that it would kill anything trying to get through the tunnel. So again we seem to be stopped.

CAUSALITY PROBLEMS

Could we get through this radiation? It's doubtful, but again let's assume we are able to build a spaceship that is so technologically advanced that it would allow us to pass through safely. Is there anything else that would give us problems? The difficulties we have talked about so far are extreme, and they are certainly sufficient to rule out space travel through black hole tunnels. But there is something even more troublesome to scientists.

These tunnels would, of course, be time "machines." If we went in one end of the tunnel, we would exit at some distant point in the universe at a different time—in the distant future or past. We could, in theory, have a tunnel that twisted around so that we exited back at Earth. If we entered this tunnel and passed through it, we would come back to the Earth many years in the future or past. Assuming that this is possible, let's consider the following scenario: you read that your grandfather was an extremely cruel person who did a lot of harm during his lifetime, and you decide to travel back in time through one of these tunnels and kill him. If this is possible, and you did kill him before he got married, you have to ask yourself where you came from. If he died before he had one of your parents, how did you come to exist? This is known as the grandfather paradox and it's obviously a serious difficulty.

Let's look at the problem from another point of view. Assume you went into a time tunnel at 11:00 AM passed through it and returned to roughly the same point on Earth at 10:55 AM of the same day. If so, you would see yourself getting into your spaceship to leave. You could, in fact, run over to yourself and say "hello." Indeed, you could even stop yourself from going into the tunnel. It sounds crazy, but if time travel into the past is possible, we have to face problems like this. In fact, in the above example, if you actually went back in time and saw yourself leave, you have to ask:

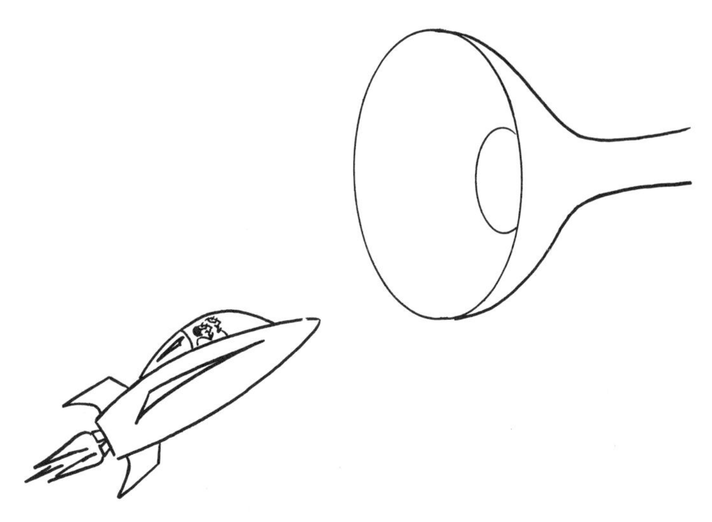

"Do you think we should chance it?"

Why didn't you see yourself watching you leave the first time. After all, it was just a replay of what happened earlier.

There are many mind-boggling paradoxes of this type that you can dream up. They all violate what is called the causality principle. Simply stated, this principle says that the effect of something always comes after its cause. If a pitcher throws a baseball across the plate and a batter hits it, it goes against common sense to say that the ball was hit before the pitcher delivered it. The idea that effect can come before cause is irrational to the human mind.

As strange as it may seem, though, scientists are seriously asking themselves if it is, in principle, possible to violate causality. Somehow it seems that if time travel is possible, it should be possible to violate causality. If, in fact, causality cannot be violated, would time travel into the past be prohibited? At the present time we don't know for sure. There are many ways we could travel into the past and not violate causality. It may be possible for us to travel into the past, if we don't interfere with it in any way. But would this really be time travel? After all, we can, in a sense, do that now by just looking at an old movie of ourselves.

Another suggestion that has been made by a number of people is based on what is called the "many world interpretation of quantum mechanics." In 1957, Hugh Everett of Princeton University suggested that it was possible for the universe to "branch off" into many universes, according to all of the choices that are possible at a given event. We refer to these other universes as parallel universes. This means that if something different happened in the past (e.g., such as you shooting your grandfather), it would branch off into a different, parallel universe—one that you were not familiar with. Most scientists no longer take this possibility seriously, but it has been talked about extensively.

I asked several scientists working in the area if they thought causality could be violated. "It's an interesting question. People have just assumed that we can't because of the grandfather paradox," said Don Page of the University of Pennsylvania, who has spent many years working on black holes and cosmology. "I think the causality problem should be investigated," said Page. "I'm not sure we can rule out the possibility of going back in time because of it. I'm not a hundred percent sure that causality in the usual sense has to be satisfied. There may, for example, be a consistency condition on the whole of space, but if you only look at a small piece of space, causality can be violated." He paused briefly. "Whether or not the universe actually has a property like this is, of course, another problem." He went on to say that you can look at the problem from two different points of view: classically and quantum mechanically. He felt that it was a much more complicated problem quantum mechanically, but in the long run we would probably have to apply quantum mechanics.

Richard Price of the University of Utah is less convinced that causality can be violated. "It seems to me that [causality violation] is an important reason to think that there are problems with time travel . . . and that it isn't possible to travel back in time." John

Wheeler replied to the question with, "I wouldn't bet any money on it. No, I don't think causality can be violated." And Michael Morris of the University of Waterloo said, "Yes . . . it's pretty disturbing that you've got this problem with time tunnels. What you have to ask is: What is going wrong? Is this a sign that the whole idea of wormholes [tunnels] is incorrect?" He went on to say, however, that he isn't convinced yet that time tunnels can be ruled out.

All in all, though, it is obvious that causality poses a serious problem for time tunnels.

WORMHOLES

So far, we have been talking about the space-time tunnels associated with black holes. These tunnels do exist according to general relativity. But there is another possibility. Schwarzschild's solution of Einstein's equations showed that a "wormhole" or tunnel in space-time is possible, independent of a black hole. We have no idea how or if these wormholes would form. But, let's ignore this problem for now and assume that they do exist. We'll refer to them as Schwarzschild wormholes.

Assuming that they exist we ask: Is it possible to use them for space travel? Again, using Einstein's theory, we can look into this possibility. The conclusion that has been reached is that, in general, they are not much better than black hole tunnels. The gravitational tidal force at the throat of a Schwarzschild wormhole is roughly of the same strength that it is at the event horizon of a Schwarzschild black hole. Furthermore, like black hole tunnels, they are unstable and tend to pinch off when you enter them. The problems are perhaps not as severe as those associated with black hole tunnels, but they are serious enough to make it unlikely that we will ever be able to use them.

On the basis of what we have seen, it appears as if it is virtually impossible to use either black hole tunnels or Schwarzschild wormholes for space travel to distant points of the universe. Of course, there's a lot we don't know about black holes and it is possible that one day we will see a way around the difficulties. But, based on what we now know, we have to rule them out for space travel. Does this mean that we will never be able to travel to the stars? In the next chapter we'll see that there may be a way.

CHAPTER 11

Overcoming the Problems

The problems are so overwhelming that it is quite unlikely that space-time tunnels associated with black holes will ever be used for interstellar travel. Even Schwarzschild wormholes appear to be of little use. Because of the difficulties, scientists had generally given up on the idea by the early 1970s. Science fiction writers, on the other hand, continued to use them, perhaps because they provided a handy way to move their characters around in the universe. From this point of view, I suppose you can't blame them. But relativists knew that what they were suggesting was impossible, and to some degree it irritated them. Why should science fiction writers continue to advocate something that had been proved impossible. It seemed crazy.

SAGAN AND THORNE

But if there was a better way, or any way for that matter, to get to the stars, it was up to the scientists to find it. In general, though, such questions are outside the research interests of scientists—in fact, they usually tend to avoid them because of the bad publicity it sometimes brings. In the mid 1980s, however, a solution to the problem was found—almost by accident. A few years earlier, Carl Sagan of Cornell University had begun writing his science fiction novel *Contact*. In the novel, his heroine Ellie Arroway detected a radio signal from the vicinity of the star Vega. With some help she was able to break the code and translate the message. It was the building instructions for a "machine," a machine that eventually took Ellie and several of her colleagues to a planet near Vega.

In the original version of the book, Sagan used the tunnel associated with a black hole to get them to Vega. But, as a scientist, he wanted to be sure that the science in the novel was as accurate as possible, so he sent the manuscript to one of the foremost black hole experts in the world, Kip Thorne of Caltech, and asked him to look it over.

Thorne was disturbed when he saw that Sagan had used a black hole. He knew there were many problems associated with them. He was also well aware of the difficulties with Schwarzschild wormholes. Still, he wanted to help Sagan. So he wrote down Einstein's field equations and began looking for a solution that had been overlooked—a solution that might allow time travel. And he found one. It was so simple he wondered why it hadn't been noticed before. Einstein's equations predicted that you could use a wormhole, not one associated with a black hole, but one that existed by itself, and you could get around the pinch-off effect. He sketched out the solution and left one of his students to fill in the details.

WORKING OUT THE DETAILS

The student that Thorne gave the problem to was Michael Morris. Born in Indianapolis in 1960, Morris did his undergraduate work at Purdue. Upon graduation he went to England for a year to work in radio astronomy. His major interests, however, were theoretical rather than observational and as a result he ended up working on a theoretical problem: an investigation of the acceleration of particles in a supernova explosion. "It was a year of fiddling around with something I wasn't sure I was interested in," he said, referring to the work.

When he returned to the United States he went to Caltech,

Kip Thorne.(Courtesy William R. Kenan, Jr.)

still not sure what he wanted to do. After about a year, however, he was attracted to the work that Kip Thorne and his group were doing and decided to join them.

Thorne gave Morris the problem and told him to work out the details. Thorne had established that wormholes were the key, but ordinary Schwarzschild wormholes wouldn't do; they had to have special properties. First, a solution had to be found that didn't pinch off.

Pinch-off was not the only problem, however. To be traversable the wormhole had to have several other properties. Thorne and Morris listed them:

1. They had to have small tidal forces.
2. They had to be two-way, which meant that they could not have a horizon.
3. Transit times through them had to be reasonable, both from the points of view of the traveler and the people outside the tunnel.
4. Radiation effects had to be minimal.
5. You had to be able to construct the wormhole with reasonable materials and within a reasonable period of time.

The most serious problem stopping them was from pinching off. Thorne had come to the conclusion that some sort of matter would have to be threaded through the wormhole—matter that would produce an outward pressure strong enough to contain the inward squeezing of the wormhole. He found a solution of Einstein's field equations that allowed him to do this. But there was still the problem of making the solution consistent with the requirements on their list.

They soon found that all the problems on their list could be overcome. In many ways, it was like a miracle. The new solution showed that tidal forces could be made as weak as those on Earth. Transit times between the ends of the tunnel could be kept to less than one year. Furthermore, the wormholes did not have to be overly large. Earlier, I mentioned that in the case of black hole tunnels, the only way around huge tidal forces was to use massive black holes with tunnels that were thousands of miles across. These tunnels could, in theory, be smaller than a mile across.

The only real problem was the material that was needed to thread the tunnels. It couldn't be ordinary matter. In fact it's not an overstatement to say that it had to be very extraordinary, even "exotic."

EXOTIC MATTER

The inward pressure on the tunnel would be so great that the matter needed to counteract it had to exert an outward tension comparable in magnitude to the pressure at the center of a neutron star. This meant the magnitude of the tension would have to be greater than the total density of mass energy of the wormhole itself, which, in turn, meant that it might have negative mass (negative mass implies negative gravity). There is nothing known on Earth that is even remotely like this.

If the discovery had been made a few years earlier, Thorne and Morris would likely have shelved it immediately. During the 1960s and 1970s, negative-energy densities were unthinkable. There were, in fact, theorems stating that it could not exist. Then, in the mid 1970s, an important breakthrough was made. Hawking showed that black holes can evaporate, which in turn, meant that the event horizon had to shrink. This was forbidden by classical theory, but Hawking showed that it was possible quantum mechanically. The important point, though, is that negative energies are involved. Just outside the event horizon, particle pairs are created out of the vacuum. They cause a flow of negative energy into the horizon that causes it to shrink.

This, however, is not the only evidence that negative energy can exist. As long ago as 1948 the Dutch physicist Hendrik Casimir showed that the energies in a field between two conducting metal plates can be less than zero. In a vacuum, particles are continually appearing and disappearing so fast that they can't be observed. The total average energy of these particles is zero. But if two conducting plates are placed in the vacuum, the energy between the plates is lowered. Since it starts at zero, it has to become negative. This means that, in theory, you could place a round metal plate over each of the mouths of a wormhole and produce enough negative energy to stop it from collapsing.

These and other arguments convinced Thorne and Morris that exotic negative-energy matter might exist, and if it did, it could be used to stabilize wormholes. Morris summarized their work for

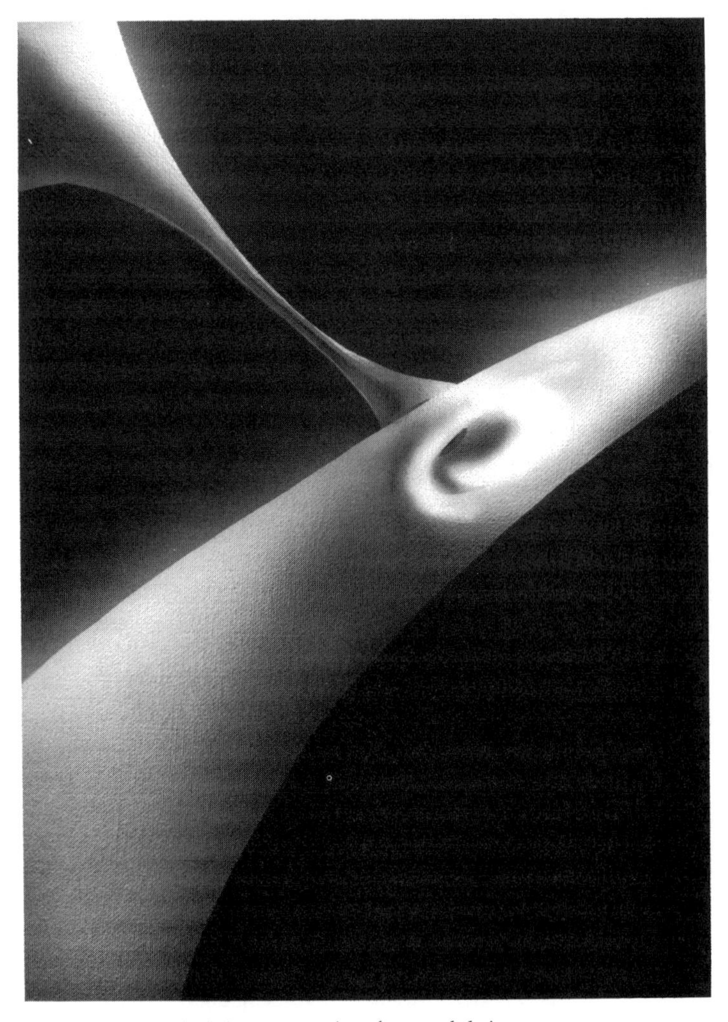

Artistic representation of a wormhole in space.

me. "We asked the question: Do the equations of general relativity allow you to have a wormhole that is everything the science fiction novelists dream of? And the answer is: Yes, but you're going to have to build it out of exotic matter . . . which may or may not exist. It's up to the particle physicists to tell us whether or not it can exist."

Thorne informed Sagan of their discovery. He told him that the tunnel would have to be threaded with exotic matter and it could be done in one of two ways. The tunnel could be lined with exotic material, but in this case the travelers would have to be protected from it. For this he suggested that a "vacuum tube" could be stripped down the tunnel. A second possibility, he said, was propping up the tunnel by using exotic matter throughout it, but choosing exotic material that passed through the human body. Neutrinos are an example of particles that do this. The density of these particles would have to be incredibly high, but anyone in the tunnel would not notice them. Sagan selected the first of the two options for his novel.

It appears, however, that exotic matter would pose other difficulties. If it had negative mass, it would also have negative gravity, or antigravity. This means that instead of being attracted to it you would be repelled; since the repulsion is proportional to the mass, which is exceedingly large, it seems that the exotic matter might create a serious problem. I asked Morris about this. "Antigravitational effects wouldn't be a problem for anyone passing through the wormhole . . . as far as I know," he said. "But you might worry about other effects it could cause. You could imagine some weird effects if you had globs of this stuff hanging around . . . but it's not clear that it would violate anything." Because of this possibility, Morris and Thorne have suggested that the exotic material can be kept to a minimum in the tunnel. They suggest that it be restricted to the central region around the throat, and be surrounded by ordinary matter.

Shortly after Morris and Thorne showed that wormholes might be made traversable, Matt Visser of Washington University in St. Louis discovered a theoretical wormhole that employed a minimum of exotic material. He spliced together two identical regions of flat space, much as we might take two cone-shaped paper cups, cut a small section off the narrow ends, then join them together at the cuts. He found that exotic matter is needed only in the curved regions of the tunnel. Therefore, if you selected the joining regions to be in the form of a cube or a polygon, the exotic matter could be confined to the corners and edges. One could easily pass through the space within the cube or polygon without any ill effects.

TRANSFORMING THE WORMHOLE INTO A TIME MACHINE

So far, we have concentrated on stabilizing the wormhole and reducing the tidal forces. We have said nothing about how it affects time. In essence, we still have to make it into a time machine.

The germ of the idea that was used for doing this came from Thomas Roman, a theorist at Central Connecticut State University. At a conference in 1986, he talked to Morris and learned about his work on wormholes. He mentioned to him that it should be possible to transform wormholes into time machines using special relativity. Morris didn't think this was possible at first, but the more he thought about it, the more he realized it might work. Looking into the details, he found that someone entering two wormholes, one after the other, could end up traveling back in time. He showed his calculations to Thorne and together they refined the idea.

They found that they could transform a single wormhole into a time machine that could take them into the past and future, depending on which direction they traveled through it. To understand what they did, let's go back to special relativity for a moment. According to this theory, if one clock moves relative to another with constant speed, it will click off seconds at a slower rate, as seen by an observer at rest near the second clock. This led

Spaceship entering a wormhole.

to what is called the twin paradox (we discussed this earlier). If one twin sets off into space at a speed close to that of light relative to his brother, he will be younger than him when he returns, with the difference depending on how fast he traveled.

Thorne and Morris, together with Ulvi Yurtsever, another of Thorne's graduate students, found that they could do something similar with the two mouths of a wormhole. If they moved one of the mouths off into space at high speed, then returned it to a point near the other mouth, they found that the wormhole would become a time machine.

To understand how this is possible, you have to consider clocks inside and outside the mouths. All clocks inside the wormhole will record the same time, regardless of the motion of one of the ends. The reason is that the two ends are not connected by ordinary space. As we saw earlier, a wormhole can reach half way across the universe, but it doesn't thread its way through the space between the two points.

Furthermore, the clock just outside the stationary mouth has not moved relative to the one just inside and will therefore show the same time. The clock just inside the mouth that has moved, on the other hand, will show an earlier time, in the same way that a moving twin, when he returns, will be younger.

Therefore, if you go in through the mouth that has moved and come out through the stationary mouth, you will go back in time. You will exit at the second mouth before you entered the first one. In the same way, if you go in through the stationary mouth and exit through the mouth that has moved, you will go into the future.

How great is the effect? As in the case of the twin paradox, it depends on how fast the moving mouth was moved and for how long. For a significant effect, however, you have to get the mouth up to a speed of over 99 percent that of light. How could you do this? One way would be to use a heavy asteroid or something similarly massive that the mouth was attracted to. Start the mouth moving toward it, then accelerate the asteroid, keeping it just out of reach. Another way would be to put some charge on the wormhole, then use an oppositely charged object to attract it.

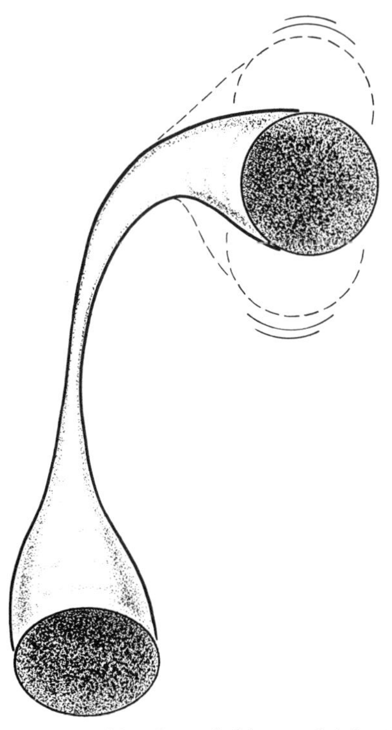

Making a wormhole into a time machine. One end of the wormhole has to be moved at a speed close to that of light.

It's important to note, though, that the clock at one mouth is just slowed down relative to the other. Because of this, you could not go further back in time than the day the time machine was created. If you wanted to go further than this, you would have to hope that wormhole time machines were created by ancient advanced civilizations—as the wormholes in Sagan's novel were.

Thorne, along with Morris and Yurtsever, published a paper in 1988 describing the conversion of a wormhole to a time machine.

They pointed out in the article that their primary objective wasn't to build one of these machines. They said they were interested only in whether the laws of physics would permit an advanced civilization to construct one.

OTHER METHODS OF MAKING TIME MACHINES

The major problem in producing a time machine as described above is accelerating one of the mouths to near the speed of light. We would need a large asteroid, which would be no problem, but bringing it close to one of the mouths of a wormhole, then dragging it through space at a speed greater than 99 percent the speed of light is something that would be difficult for even a very advanced civilization. Because of this we ask: Is there an easier way to convert a wormhole to a time machine? Valery Frolov of the Physical Institute in Moscow and Igor Novikov of the Space Research Institute in Moscow have shown that there may be a better way. Furthermore, in searching for an easier method, they came up with several other interesting results. They decided to look at what would happen if you placed a traversable wormhole in a strong gravitational field or a strong electromagnetic field. And what they found surprised them.

They began by considering what would happen if they placed one mouth of a wormhole close to the surface of a neutron star and kept the other mouth well away from it. Neutron stars, we know, are relatively common, certainly much more common than black holes. We have at least 300 good neutron star candidates compared to only a handful of good black hole candidates. They found that if one mouth was held near a neutron star, the wormhole would be transformed into a time machine. This is obviously much simpler than getting one of the mouths up near the speed of light. You would merely have to bring the wormhole into the vicinity of a neutron star and the wormhole would automatically be transformed into a time machine. Once the transformation had taken place, you could take the wormhole wherever you wanted, for example, to Earth, and use it to travel to the past.

Frolov and Novikov showed further that a strong gravitational field, such as that near a neutron star, was not necessarily needed. Even a moderate gravitational field, such as that near an ordinary star, would do, but the mouth would have to be held in the field for a much longer period of time. They then looked at what would happen during the transformation, what would take place before the conversion occurred, and they found something interesting. They discovered that the wormhole could be used as an energy generator—it could produce useful energy. You could, for example, pass photons or particles through the wormhole and have them exit with more energy than they went in with. As we saw earlier, there is a similar process in the case of black holes, the Penrose process. In that case, energy was extracted from the black hole. But where is the energy coming from in this case? It obviously has to come from the wormhole itself; as energy is extracted, the wormhole must lose an equivalent amount of energy.

Frolov and Novikov then went on to show that if the wormhole was immersed in an electric field, it could also be made into an energy generator. The easiest way to do this would be to put some charge near one of the mouths. The electric field from the charges would then thread its way through the wormhole, producing an energy generator. Charged particles could then be fed in one end and come out the other end with more energy than they went in with. Again, the energy comes from the wormhole itself.

But what would happen if we brought one mouth of a wormhole in the vicinity of a black hole, say, just outside its event horizon? Frolov and Novikov showed that you could extract the energy of the black hole. It would be like dropping a pipeline into an underground pool of oil. Radiant thermal energy would come pouring out the end of the wormhole. William Unruh and Robert Wald considered a process similar to this in 1983—they referred to it as "mining" a black hole. But they did not use a wormhole; they visualized a "box" being lowered into the black hole. Wormholes would obviously be much more convenient. With them, we would be able to tap an almost unlimited source of energy.

As amazing as this is, it wasn't the most exciting of Frolov and Novikov's results. They then considered what would happen if they lowered the wormhole right into the black hole, through the event horizon. To their surprise, they found that it would give us a route into the interior of a black hole (i.e., through the event horizon). What is most important, however, is that once inside, we would not be trapped. Because the wormhole is composed of exotic material, it is quite possible that there would be no event horizon blocking our way. You could pass right into the interior of the black hole, study it in detail, and come back out through the wormhole to your spaceship waiting in orbit. They also noted, however, that the wormhole may just push the section of the event horizon across its end closer to the center of the black hole. In this case we would not be able to pass beyond the end of the wormhole.

In the last paragraph of their paper, which appeared in *Physical Review D*, they cautioned that everything they discussed had physical significance only if wormholes do, indeed, exist and could be made into stable time machines.

TIPLER'S TIME MACHINES

Another possible but completely different type of time machine was devised by Frank Tipler of Tulane University in 1974. To understand it, it is best to go back to a discovery made in 1949 by Kurt Godel of Princeton University. He found a solution of Einstein's equations that described a rotating universe. Godel considered what we call "light cones" in this universe. We talked about them briefly in an earlier chapter. If you plot time in the vertical direction and space in the horizontal direction, scaling the axes so that the speed of light is at an angle of 45 degrees, you get a light cone when you rotate the diagram around the time axis. Your path in space-time, in other words your "world line," threads up through the bottom cone, called the past light cone, through the

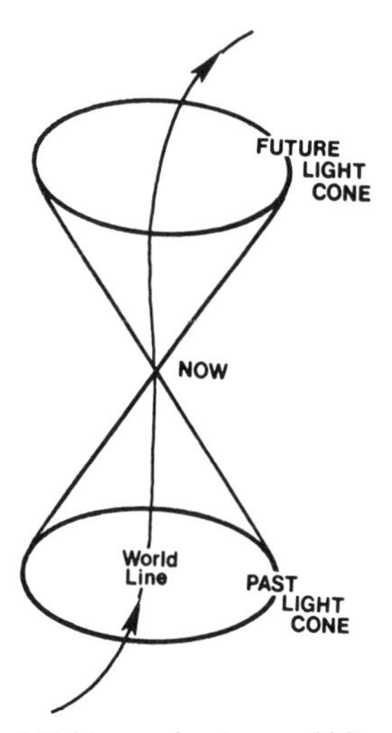

A light cone showing world line.

center, called now, and into the upper cone, called the future light cone. It is restricted to these two cones.

Godel found that the light cones in his rotating universe would become tilted in the direction of rotation to such an extent that the future light cone of one location would intersect the past light cone of another. He then showed that for circular paths perpendicular to the rotation axis, it was possible for a traveler to move into his future light cone, go around the loop, and enter his own past light cone. He would, in effect, be going around a time machine. In practice, it turns out that the circular path the traveler would have to travel was about 100 billion light-years long, so there was little serious interest in the discovery.

In 1974, however, Frank Tipler picked up on the idea and showed that you didn't need the entire universe to rotate. He determined that if you rotated an infinitely long massive cylinder

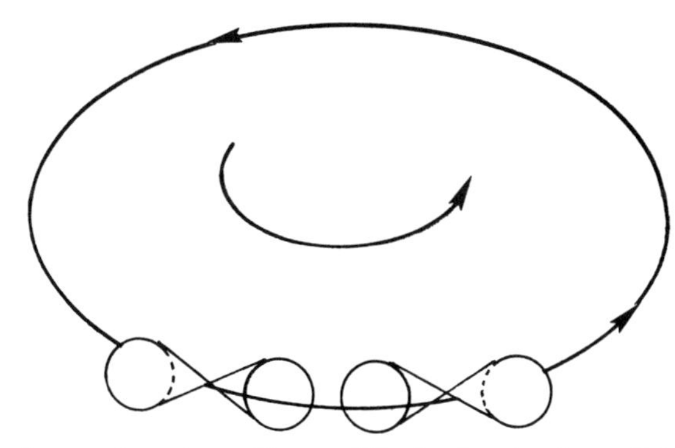

Light cones in a rotating universe. They would be tipped so that the past light cone of one location intersects the future light cone of another. This could be used to make a time machine.

fast enough, it would also tip light cones and could be made into a time machine. The speed at the outer surface of the cylinder, however, would have to be greater than half the speed of light.

Science writer John Gribbon then showed that a cylinder 100 km long with a 10-km radius and a mass roughly equal to that of our sun could be made into this type of time machine if it were rotated at a speed greater than half the speed of light. In fact, he noted that you don't really need a cylinder. An asteroid of roughly this size would do.

While this sounds as if it may be possible to do, Tipler points out that there are many problems. He has shown, for example, that if you could get something rotating this fast, part of it would likely collapse into a singularity. Furthermore, the amount of energy required to send physical objects through it would be incredible. Tipler believes that such a machine might be useful for sending messages into the past, but not physical objects.

CAUSALITY REVISITED

In the preceding chapter, we saw that if time machines actually could be constructed, we would have a serious problem with causality. And there appear to be no laws of physics that forbid them. Because of this, much of the recent work in the area has been directed toward resolving the conflict between causality and time travel, or at least proving once and for all that, because of causality, time travel is impossible.

Kip Thorne and his group at Caltech are trying to see if there is any way you can travel back in time and not violate causality. They've been looking at the problem by considering a "billiard ball" version of it; they believe they can demonstrate a case where there is consistent physics. Assume we have a wormhole that has been set up as a time machine. The wormhole will, of course, have two mouths; let's call them A and B. Suppose now that a billiard ball enters mouth A, then comes out of mouth B at a slightly earlier time. If, when it comes out of B, it hits itself on the way into A and knocks itself so it can't enter A, we have an inconsistent trajectory. Something's obviously wrong. The billiard ball entered A, yet it didn't.

If this actually happened, you'd have to say that we didn't have a consistent description of physics. To get around this, you have to have a billiard ball trajectory that gives proper physics. For example, you have to have a billiard ball in the region outside of mouth A moving in the same way it did earlier, but, before it enters A, it has to hit a ball coming out of mouth B. But it can't hit it head-on so that it stops it from going in. It has to hit it a glancing blow that deflects it slightly, but still allows it to enter mouth A. Then you have to adjust the angle at which the ball enters A to make the angle at which it comes out of B just right as to allow a self-consistent trajectory.

This may sound like a silly game to some, but it is taken quite seriously and it may have serious implications for physics. If it can be shown that there are definitely no laws forbidding time travel into the past, there's no question but that we have to resolve the

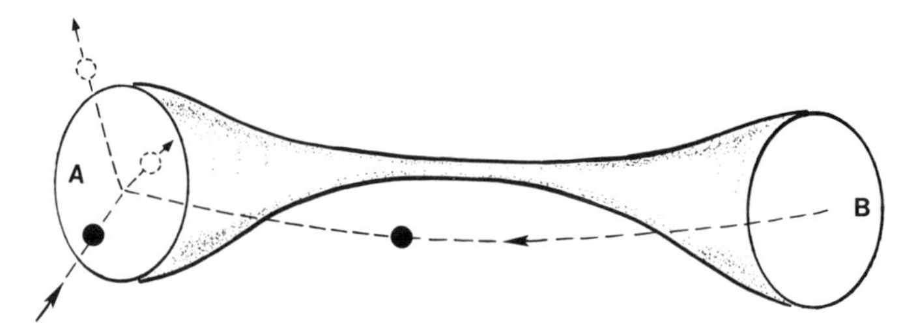

Testing causality. A billiard ball comes out of mouth B and strikes itself going into A.

causality paradox. "If you view this stuff in the right way, it's not crackpot," said Richard Price of the University of Utah. "It's the search for what is in physics that prevents, or doesn't prevent, time travel." Still, he admits he feels a little uneasy about the topic. "I'm not fascinated by these questions myself," he said. "I think its a point of interest . . . but, as far as I'm concerned, I think time travel has received more attention and notoriety in the popular press than is appropriate."

FRIEDMAN

Several of the other scientists that I talked to also felt a little uneasy about the subject of time travel. John Friedman of University of Wisconsin, for example, quipped, "I also do things that are closer to reality," referring to his work on things other than wormholes and time machines. He still spends some of his time working on rapidly rotating neutron stars, but much of his time in recent years has been spent working on time travel. Born and raised in Chicago, Friedman took his undergraduate work at Harvard and got his Ph.D. at the University of Chicago, working under Subrahmanyan Chandrasekhar on a problem in astrophysics.

Friedman got involved with traversable wormholes indirectly. He was interested in quantum gravity and was trying, theoretically, to model elementary particles out of space-time topologies that resembled wormholes. Then Michael Morris came to the University of Wisconsin on a one-year postdoctoral fellowship, and they began working together. Most of their work was concerned with "closed timelike curves," in other words, closed curves that allow passage back in time. Along with Morris he looked at a problem similar to the billiard ball problem Thorne and his group were working on. "We looked at a simple version of this problem," he said. "We asked: Suppose you have a space-time and a closed timelike curve. Can you do proper physics in that space-time? Can you, for example, impose initial conditions for a simple case, and find a unique evolution in time." Friedman admits that so far he has not been able to rule out time machines.

BUILDING A TRAVERSABLE WORMHOLE

So far we've been talking about traversable wormholes as if we could easily make one. But could we? Carl Sagan got around this problem in his book *Contact* by saying that they were constructed long ago by an ancient, extinct civilization as part of a rapid galactic transit system. But we need a better answer than this. Morris, Thorne, and Yurtsever do make a suggestion in their paper on wormholes. "One can imagine an advanced civilization pulling a wormhole out of the quantum foam and enlarging it to classical size," they write.

The quantum foam they are referring to is a foam of space and nonspace, similar perhaps to soap foam—a fluctuating region filled with wormholes that exists on a scale much smaller than atoms. Although it is not postulated directly by any theory, it seems to be a natural outgrowth of what is known about quantum gravity. It was first suggested (at least a version of it) by John Wheeler in the mid 1950s. I asked Wheeler if he is still convinced that it exists. He said,

> I think it's inescapable. I would say of all the advances in theoretical physics in the immediate postwar period the

> number one advance was the proof of the existence of fluctuations in the electromagnetic field. And the arguments that make these microscopic fluctuations inescapable also apply to the geometry of space—what I call geometrodynamics. The only thing is that they are more dramatic in their nature in the case of space, and less impressive in their consequences for everyday atoms. That is to say, the quantum fluctuations in the geometry of space-time are only seen when you get down to the Planck length [10^{-33} cm].

Wheeler told me about the day the idea came to him. "I was giving a lecture in Leiden, when it suddenly occurred to me that wormholes are not necessarily a property of particles, they're a property of space. A student that was with me at the time went down and had an artist draw a picture of them for me to illustrate to the class."

Virtually all physicists are now convinced that this quantum foam exists, and that it is filled with little wormholes. "It's certainly plausible, that at these scales, there are all kinds of little wormholes and weird space-time curvatures connecting, separating, and recombining all the time. The question is: Is there any law of physics that prevents you from starting with a tiny wormhole and making it larger? I don't know," said Morris.

It would, of course, be a considerable feat of engineering if we could pull them up out of the quantum foam. So far, though, there are no other suggestions on how we might create them. I asked Don Page, who is now on leave at the University of Alberta, what he thought of the suggestion. Page has been working recently with Hawking on wormholes. "I'm sure it's just conjecture," he said. "But I can't give you any evidence that it's impossible. We certainly don't know of any method of bringing them up at the present time. But if you could solve the problem of producing negative energy densities, I suppose it might be possible." John Friedman is of the same opinion. "The question of whether you could enlarge them to get macroscopic tunnels is very close to the question of whether you could maintain them after you got them enlarged, whether in trying to enlarge them you end up violating the average weak energy condition. If I had to bet, though, I'd bet against being able to enlarge them and maintain them. But it hasn't been ruled out yet."

QUANTUM GRAVITY

One of the major roadblocks in our understanding of the quantum foam is that we don't have an adequate theory of it. It seems that it has to exist, scientists see no way around it, yet they cannot adequately describe it or make proper calculations of its properties. When we get down to the realm of atoms and elementary particles—the quantum realm—general relativity breaks down. We therefore need a quantized version of general relativity. But there are many problems in trying to construct such a theory. Ian Redmount, a former student of Thorne, who is now at Washington University in St. Louis, pointed out one of the difficulties to me. "The real problem is that we have no quantum gravity phenomenon," he said. "So it isn't clear what a quantum gravity theory is supposed to say—much less, how it's supposed to say it. People therefore do things formally. They take what works from quantum field theory and generalize it to the case of the gravitational field . . . but in doing this you get all sorts of problems."

A quantum theory of gravity is also expected to be crucial to our understanding of the earliest moments of the universe. Going back in time to the big bang singularity, we find that, for the first fraction of a second, the universe was exceedingly hot and dense. At 10^{-43} second, for example, it had a density of 10^{94} grams per cubic centimeter. This epoch is referred to as the Planck era and is the realm of quantum cosmology. Theorists conjecture that space and time were a discontinuous foam at this time. Time, as we know it today, did not exist.

But if this foam of wormholes existed in the very early universe, and the universe expanded, why aren't there wormholes all around us now? I asked Redmount about this. "Anything that is bound, gravitationally or otherwise, does not expand," he

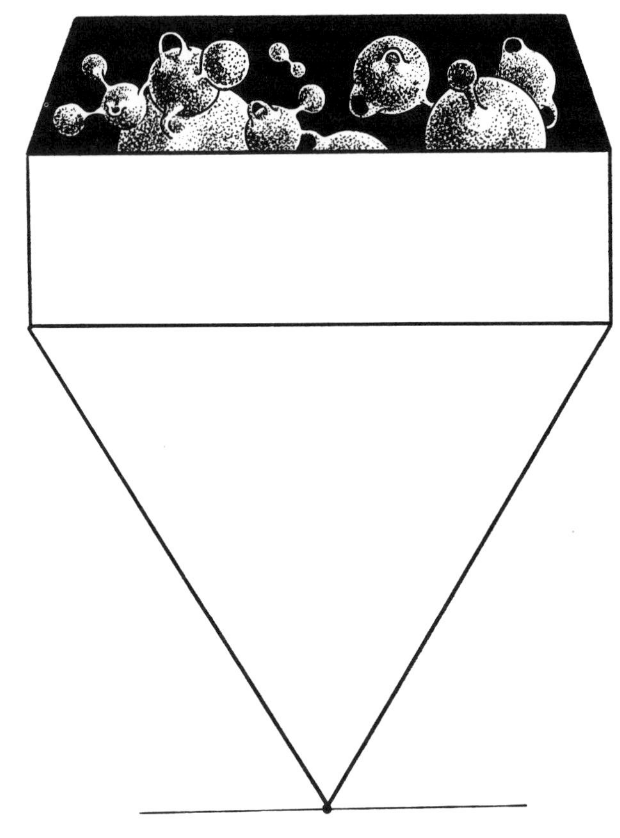

A simple representation of the quantum foam. This shows what a magnified view might look like.

said. "And the binding forces of the wormhole would be stronger than its tendency to expand." He paused. "I suppose you could imagine a model in which the wormholes got bigger and were spread far apart . . . although certainly the big bang model doesn't predict it. But if this happened, you would expect the distance scale for these things to be approximately the same as it is for any other cosmological object. The nearest wormhole might not be any closer than the nearest quasar. In that case, using them for space travel would be impossible."

Scientists have been struggling for several decades to quantize general relativity. Today, however, most efforts are going into an even more encompassing theory, a unified theory of all nature. This theory would include quantum gravity, but it would also explain the plethora of particles and fields. It would, in essence, explain all of physics.

In recent years, there has been some progress in this direction. The most promising line of research is what is called string theory. In 1974, Joel Scherk of the University of Paris and John Schwarz of Caltech showed that string theory can describe the gravitational field. But few scientists paid much attention to the work until 1984 when John Schwarz and Michael Green showed that they were able to "build" elementary particles out of tiny strings, and that these particles had many of the properties that we observe. Within a short time, many people were working in the field.

We are still not certain that string theory, or the newer superstring theory, will give us a quantized theory of gravity or a unified field theory, but it does appear to be a step in the right direction. I asked John Wheeler for his thoughts on the possibility of achieving a unified theory. "I think we're going to have to go to a whole new way of looking at things before we can see our way through this business. My motto is: The quantum is the 'crack' in the armor that covers the secret of existence. And I think we have to find out what lies under that armor to get to the bottom of things. That's the biggest mystery." I asked him if he thought superstring theory would give us any of the answers. "It's a step on the way, but to my mind no step will be truly there until we work our way out of the presupposition that there is any such thing as time and space."

Richard Price thought that a quantized theory of gravity would definitely eliminate a lot of the speculation. "But even if we had such a theory," he said, "I believe there would still be unanswered questions. It would clarify things, but it wouldn't answer all the questions." John Friedman is worried that even if we had such a theory it might be extremely difficult to perform the relevant calculations.

Morris is a little more optimistic. "We'd be able to say a lot more things about these ideas [about the quantum foam] if we had such a theory," he said. "As it is, though, there is quite a bit we can say from what we know already." And finally, Redmount commented, "Hawking is convinced we'll have such a theory by the end of the century. Maybe he's right. But I'm less confident."

CHAPTER 12

Wormholes and Other Universes

According to a well-known theorem in classical black hole physics, you need a singularity to produce a wormhole. On the quantum level, however, space is filled with tiny wormholes. It appears, then, that if we are ever to have a large wormhole, we are going to have to start with a small one and enlarge it. We have no idea at the present time how we would do this, but we do know that it is important to try to understand how these tiny quantum wormholes behave. As it turns out, there has been a considerable amount of work in this area in the past few years.

HAWKING'S WORMHOLES

Hawking was one of the first to take wormholes seriously. His interest was sparked a few years after he discovered that black holes evaporate. As we saw earlier, a black hole has a surface temperature. This temperature is extremely low for large black holes, but can be very high for tiny ones. Furthermore, any object with a temperature greater than zero degrees absolute radiates, hence black holes radiate. As a result, they lose mass and grow smaller, but as they grow smaller, they radiate more intensely, and this in turn makes them grow smaller even faster, until finally, in the last moments of their life, they explode and disappear.

What is left after the explosion? For a number of years scientists speculated that there might be a naked singularity, a singularity with no event horizon around it. Hawking is now convinced, however, that the black hole completely disappears, singularity and all. It is lost to our universe. This means that everything that ever fell into the black hole is also lost.

But to say that it just disappears isn't really satisfactory. It obviously has to go somewhere. And Hawking believes he has an answer for this, although it is one that you may have a hard time believing. He is convinced that it goes into another universe, a universe that branches off ours. The remains of the black hole, he says, tunnel via a wormhole to this other universe.

Hawking is quick to point out, though, that this is prohibited in classical general relativity, but he gets around it by using quantum gravity. Basically, what this means is that it is prohibited on a large scale, but allowed on a quantum scale. I mentioned earlier, though, that we don't have a quantum theory of gravity. True, but a few years ago Hawking introduced a theory that describes the universe through the use of a quantum wave function; he refers to it as his "wave function of the universe." There are many problems with this theory, but Hawking believes that it may be a good approximation to a quantum theory of gravity. Furthermore, it is possible, using this theory, to perform quantum gravity calculations. We're not certain how good they are, but they're better than nothing.

Hawking's quantum theory of gravity predicts that there are tiny wormholes popping in and out of existence continuously throughout space. The scale at which they do this, however, is incredibly small (Planck size, 10^{-33} cm). These wormholes have two mouths: one is in our universe and the other sometimes in our universe, sometimes in another universe. In some cases, it also ends in a tiny "baby universe," a universe that could, in theory, inflate to a large universe. It's not much use asking where these other universes are relative to ours. The question makes no sense, since they aren't separated by space in the usual sense. The easiest way to think of these universes is as a cluster of balloons, where the region between the balloons is filled with "nothing."

In one respect, though, Hawking's theory is quite different

Close-up view of baby universe and tiny wormholes.

from general relativity. In general relativity we have three space coordinates and one coordinate of time. All are real. In Hawking's theory, on the other hand, time is imaginary (the square root of an imaginary number). Some physicists are uneasy about this, and feel that it is not right. With his theory, however, Hawking has shown that wormholes exist, and when a black hole evaporates, it disappears from our universe by tunneling through one of these wormholes to another universe.

Hawking has also used his wormholes to explain particle interactions. In the case of the mutual annihilation of an electron and its antiparticle (the positron), for example, he visualizes both particles falling into a wormhole, with a photon coming out. The wormhole is so small, though, that it just appears that an electron and a positron have come together and annihilated, with the release of a photon. We don't see the role the wormhole played. But having the interaction take place in a wormhole is advantageous; it has allowed Hawking to explain things about these interactions that can as yet be explained in no other way.

Most physicists were not bothered by Hawking's "other universes," but they were troubled by something else he suggested. He believed that when the particles from our universe passed through these wormholes, the large-scale coherence of our universe was destroyed. According to his calculations, this resulted because there was a "loss of information" (information about values of fundamental constants, for example) through these wormholes. This, in itself, might not seem serious, but Hawking said that if it was true, we were wasting our time searching for a unified theory of the universe. Such a theory would be useless.

CONTROVERSY

The problem of loss of information actually goes back farther than the discovery of wormholes. There was a similar problem associated with black holes that was also brought to our attention by Hawking. In 1976, he published a paper titled, "Breakdown of

Predictability in Gravitational Collapse." In this paper, he pointed out that the classical concepts of space and time along with the laws of physics break down in the singularity of a black hole. But he claimed that the breakdown was not due to our ignorance of a better, more correct theory, rather it was a "fundamental limitation in our ability to predict the future." He was convinced that there was an uncertainty in relation to black holes which was analogous to the uncertainty that exists at the atomic level in quantum mechanics. He referred to it as the "principle of ignorance."

The principle of uncertainty in quantum mechanics is a fundamental part of quantum theory. It tells us that we cannot simultaneously measure two variables such as momentum and position with precise accuracy. The more exactly we measure one, the more inexact, or "fuzzier" the other becomes. Einstein did not like this aspect of quantum mechanics and was convinced that it was wrong. "God does not play dice," he said. Hawking countered this with, "God not only plays dice, He sometimes throws the dice where they cannot be seen."

Hawking's principle of ignorance threw the world of physics into turmoil. Many scientists were, at this time, looking for a unified field theory. Or perhaps I should say that they were working on projects that they hoped would lead to a breakthrough in the area. But if Hawking was correct, even if such a theory were formulated, they would not be able to make meaningful calculations with it and were therefore wasting their time.

But not everyone agreed with Hawking. One who didn't was Don Page of the University of Pennsylvania, now on leave at the University of Alberta. Born in Bethel, Alaska, Page spent much of his youth in the mountains and on the lakes around Bethel. He loved to ski, fish, hike, and go dogsledding, but he also found at an early age that he had an aptitude for mathematics. He loved to prove things for himself, to dig out solutions. He read books on advanced math while still in high school and also developed an interest in cosmology after reading an article about the universe in the *Saturday Evening Post*.

Upon graduating from high school, Page went to William

Jewell College in Missouri, mostly because it was the college his parents had attended. It was a small college, but it had a good physics department, and his enthusiasm for physics grew. His teachers recognized considerable potential and encouraged him to go on to graduate school.

By the time he was a senior, he was sure he was going to become a physicist, but not sure which branch of physics he was most interested in. Particle physics intrigued him, but he also had an interest in relativity and black holes. When he graduated he went to Caltech and soon came in contact with Kip Thorne. After taking Thorne's course in relativity, he knew that this was the area he wanted to specialize in.

He hoped to do a thesis on black holes, but Thorne suggested a problem in cosmology. He had barely started working on it, however, when Hawking came to Caltech. Page got to know him and spent considerable time talking to him about relativity and black holes, and soon they were working together.

During his last year at Caltech, Page began applying for postdoctoral positions. He asked Hawking, who had gone back to England, to write several letters of reference for him. Then one day he got a letter from Hawking asking if he would be interested in a postdoctoral position at Cambridge. Page was delighted. He went to England and for the next three years worked with Hawking on black holes.

Although he worked closely with Hawking, he didn't always agree with him. One thing they didn't agree upon was loss of information into black holes. In 1979, in fact, Page wrote a paper titled "Is Black Hole Evaporation Predictable?" in which he detailed his objections to Hawking's arguments. In reply to Hawking's quote, "God not only plays dice, He sometimes throws the dice where they cannot be seen," he wrote, "If God throws the dice where they cannot be seen, they cannot affect us."

For a while, the scientific community was divided on the question of loss of information. "General relativists tended to side with Stephen," said Page, "and particle physicists tended to side with me." Then Hawking postulated the existence of "bubbles" in the vacuum. "They were basically the wormholes and baby universes of today, but we didn't call them that at the time," said Page. Within a short time, Hawking had become convinced that there was also a loss of information into these bubbles. Page and Chris Pope were working with Hawking on the project, and when they started to write up a paper for publication, a problem arose. "It was rather amusing," said Page. "Stephen was saying that these bubbles 'might' cause a loss of information from our universe, but when he wrote up his part of the paper, he said that they 'do' cause a loss of information. I had argued that they don't, so I told him that maybe we should tone things down a little. But he was reluctant—he wanted to keep the word 'do.' Chris Pope, however, agreed with me, so it was two against one. We therefore changed 'do' to 'might' in the final draft and never told Stephen."

The bubbles of this early theory eventually became "baby universes" that were connected to our universe by tiny wormholes. With their discovery, the problem of loss of information down black holes soon faded, and the new problem became the loss of information down wormholes (this is also referred to as "loss of coherence").

Page has continued working with Hawking on wormholes and black holes, and over the years has co-authored many papers with him. He is currently looking for wormhole solutions in classical general relativity, but says that he has had only limited success so far.

OTHER UNIVERSES

Hawking's announcement that there might be a loss of information down wormholes worried many people. But, as in the case of information loss down black holes, not everyone agreed with him. One who didn't was Sidney Coleman of Harvard.

Born and raised in Chicago, Coleman became interested in physics at an early age. He was about eight years old when he heard about the atomic bomb. "It terrified me, but I suppose all

Sidney Coleman.

sensible people were terrified of it," he said. "It was big news, though, and the Chicago *Tribune* ran a Sunday supplement on nuclear physics. I found it fascinating . . . but incomprehensible. So I started reading books on popular physics from the local library. It was there that I came across George Gamow's books; they had a strong influence on me. I was soon very excited about it all, and knew, even then, that I would become a scientist."

Coleman went to Illinois Institute of Technology, then on to Caltech for graduate work. His thesis advisor was the Nobel Laureate Murray Gell-Mann. "Murray was the one who signed the papers," said Coleman. "But during my last year at Caltech, Sheldon Glashow [who also later won the Nobel prize] was a postdoctoral fellow. He was my spiritual if not legal mentor. He was, in a sense, the one I really worked under, in that he followed my research carefully and gave me numerous tips."

Coleman first heard of the problem of loss of information via wormholes when he attended a lecture given by Hawking at Boston University in the spring of 1987. "During the lecture, I realized something was wrong, or rather I felt something was wrong with the way Stephen was analyzing the situation. I discussed it with him after the lecture, but didn't think of exploring it any further. The following fall, however, Steve Giddings came to Harvard as a junior fellow. At the time, he had just completed some work with Andy Strominger where they had found some wormhole solutions in the equations of imaginary-time field theory and had calculated many of the properties of these things. He was very eager to talk to me about this, but I was occupied with other things. He kept pushing me to think about it, so finally I said, 'Well, I know there is something wrong with all of this, because I had an argument I showed to Hawking.'"

Coleman reconstructed the argument he had given Hawking and showed it to Giddings. In the course of their discussion, Coleman's interest in the subject was rekindled, and he decided to investigate it further. Within a short time he had produced some amazing results; he published them in a paper titled, "Black Holes as Red Herrings: Topological Fluctuations and the Loss of Quantum Coherence." Like many of his other papers, this one also has a witty title. I asked him about this. "Well . . . why not?" he said. "They don't obscure the contents; if anything, they reveal them. But I don't really like to be 'cute.' Physics suffers from a plague of giving cute names to things [e.g., gluons, quarks]. On the other hand, you don't have to be dull."

In his paper, Coleman gave a convincing demonstration that wormholes do not destroy quantum coherence, as Hawking had suggested, and therefore there was no loss of information. But even more importantly, he pointed out that wormholes had an effect on the fundamental constants of nature—such things as mass and charge of the electron, and the gravitational constant (the constant of nature that determines the gravitational field). As strange as it may seem, we have no idea why these constants have the particular values they have. Coleman hoped he could explain them. Giddings and a colleague, Andrew Strominger, published a paper back-to-back with Coleman's (in the same issue of *Nuclear Physics*) that came to roughly the same conclusion.

Shortly before this paper was written, Coleman had attended some lectures given by Andrei Linde of the USSR, who had come to Harvard to talk about inflation theory. "During these lectures Linde floated the idea that the cosmological constant might involve two disconnected universes communicating with one another," said Coleman. The cosmological constant is a constant that Einstein added to his equations in 1917 when he tried to apply them to the universe. He found that, without this constant, the universe was unstable—it expanded or contracted—and he didn't like this (the expansion of the universe hadn't been discovered yet), so he added the constant to stop it. Later, however, he rejected it, referring to it as the "greatest blunder of his life."

But others didn't cast the constant off so easily. In fact, it appears to play an important role in modern quantum field theory. It is well-known that the vacuum is permeated with short-lived, virtual particles, and because of this the vacuum has a high-energy density. The cosmological constant in Einstein's equations gives a measure of this energy. The constant accounts for this energy, and as a result, quantum field theorists prefer to keep it. But even with the constant there is an enigma. There are many different types of particles and antiparticles generated in the vacuum. Some of them make a positive contribution to the cosmological constant, while others make a negative contribution. Strangely, however, the positive contribution exactly cancels the negative contribution. This is particularly amazing in that the contribution from each side is incredibly high, of the order of 10^{80} energy units.

This is a puzzle that has plagued cosmologists for years. But thinking about Linde's talk, Coleman realized that he might have a way of explaining it. "In my mind I made a connection that the things I had found for wormholes might have an application to this problem," said Coleman. "I talked to a lot of people around Harvard about it, but I just kept going around in circles and getting nowhere. Then suddenly it came to me. It was something I had heard Hawking say in a lecture several years earlier. I understood very little of the lecture at the time, but in beating my head against this problem I remembered it, and realized it would apply here. The whole thing came together quite quickly after that." In the spring of 1988, he published a second paper titled, "Why There is Nothing Rather Than Something: A Theory of the Cosmological Constant." In this paper, he argued that quantum tunneling between universes via wormholes made the cosmological constant vanish. Thomas Banks of the University of California came to the same conclusion at roughly the same time.

The papers threw the world of cosmology into an uproar. According to Coleman's (and Bank's) theory, there are a large number of other universes around like the one we live in, but they're all connected by tiny microscopic wormholes. These wormholes affect our universe in that they allow information from these other universes to flow into our universe, and this, in turn, affects our universe's fundamental constants. Because of these wormholes there is only a certain "probability" of having a specific value for a fundamental constant. These constants don't have explicit, well-defined values.

The idea that there may be other large universes like ours may sound crazy, but it's not new. People have argued for years that this might be true, but in most cases it was pure speculation, with nothing to back it up. Such universes are sometimes called parallel universes, but Coleman doesn't like this word. "They're not really parallel universes," he said, "I never used that word. They're large 'other' universes on the other side of the wormhole. But they aren't universes we can travel to." He paused. "It's hard to explain them without getting awfully technical. Everything's taking place in imaginary time." He paused again. "I hate to use those words [imaginary time] because they suggests bafflement and confusion . . . which isn't really true. Anyway, it's not taking place in real time. So these other universes are no more parallel universes than

"I've been hunting for those 'other universes' for years, and so far I haven't seen a sign of them."

the Andromeda galaxy is. The Andromeda galaxy is just 'another place' and the difference between the 'other places' that appear in my theory and the Andromeda galaxy is that we can, in principle, go to the Andromeda galaxy, but we can't go to the other places in my theory."

In an article on baby universes, Hawking also comments on this question, ". . . [these wormholes] would not be much good for taking short cuts to other galaxies. Not, that is, unless you can move in imaginary time."

Another possibility is that we could use these wormholes to communicate with other universes, in the way we communicate with one another via the telephone. Coleman said that this is also impossible. All we would ever get (assuming we could get anything) would be garbled messages from many different universes.

OTHER WORMHOLES

As I mentioned earlier, just before Coleman wrote his first paper, Steve Giddings and Andy Strominger discovered wormhole solutions in a theory they were investigating. Born and raised in the Salt Lake Valley, Utah, Giddings was encouraged to go into science by his father, a chemistry professor at the University of Utah. "He told me about the mysteries of quantum mechanics and it really amazed me," said Giddings. A TV program called "Key to the Universe" also had a strong influence on him, he said. As a result of it and talks with his father he began reading books on quantum theory. "One that I particularly remember," he said, "was *The Strange Story of the Quantum*."

Giddings took his undergraduate work at the University of Utah, then went to Princeton University for graduate work. At Princeton he met Andy Strominger who was at the nearby Institute for Advanced Study. Strominger studied at Harvard as an undergraduate, then went on to MIT for graduate work, and is currently working at the University of California at Santa Barbara.

"At one point I thought of asking Andy to be my thesis advisor," said Giddings. He didn't, but he continued to talk to him about quantum gravity and string theory, and soon they were working together.

Giddings, who is now also at University of California at Santa Barbara, and Strominger found their wormhole solution while examining the equations of string theory. They explored the solution, trying to figure out what observable effects it might have. Then they heard that Hawking had found something similar

Steve Giddings.

during the summer of 1987, and Giddings started to talk to Coleman about the discovery.

"We came upon this solution in the spring of 1987 and realized it was interesting," said Giddings. "But instead of publishing it right away, we spent the summer trying to figure out how to interpret it, and what its physical effects would be." The delay was, however, useful in that it allowed them to point out that there would be no loss of quantum coherence (information), but that wormholes would affect the fundamental constants of nature.

Giddings and Strominger are still both working on wormholes. Giddings said that at the present time he is particularly interested in whether it is possible to get wormholes in real time, rather than imaginary time. "I'm looking into the question: If you start with real time, how should you go to imaginary time within the context of quantum cosmology?"

Thomas Banks, who is now at Rutgers University, also published a paper on wormholes. His main concern was their effect on the cosmological constant. It came out about the same time as Coleman's paper on the same subject, but used a different approach. In his theory time is not imaginary. He showed that the cosmological constant would be extremely small, but not necessarily zero. Just as Coleman did, he argued that because of the wormholes, there was only a "probability" that the fundamental constants of nature would have a particular value.

Born and raised in New York City, Banks credits his father, a chemist, with encouraging him to go into science. He took his undergraduate work at Reed College in Oregon, then went to MIT for his Ph.D. He said he got interested in wormholes by reading some of Hawking's papers. "For many years Hawking had been talking about this problem of loss of information in quantum gravity. I had done some work on it in 1982, trying to understand it, and found out that the way Hawking had proposed it in his original paper didn't make much sense," he said. Then, hesitating slightly, he continued. "Let me back up a little. Hawking originally talked about this loss of information in black holes, but he later said that similar things would be happening with quantum fluctuations. He never made it clear what he meant by this . . . but eventually [these fluctuations] became wormholes. According to the work I did in 1982, that didn't make much sense. But then in the late 1980s Hawkings, Giddings, Strominger, and Coleman said you could understand all of this better if there was more than one universe. That's what got me interested. I realized the old problem that I had worked on could be resolved."

TROUBLE IN WORMVILLE

Although most scientists working in the area were initially excited about Coleman's and Bank's results, controversy soon developed. And it seemed that Coleman got most of the flak. First, Igor Klebanov and Leonard Susskind of Stanford showed that Coleman's method did not give the proper mass for the pion (an elementary particle). Others showed that a similar problem existed for quarks. Then Willy Fischler of the University of Texas and Susskind argued that Coleman's theory did not restrain the size of the wormholes. According to their calculations, his theory predicted that large-scale wormholes could materialize in spacetime. And, of course, we don't see large wormholes. Finally, William Unruh of the University of British Columbia published a paper arguing that the wormholes in Coleman's theory do lead to a loss of quantum coherence; furthermore, he believes there is an instability in the theory.

I asked Coleman about these difficulties. He said that he is confident that he has successfully countered the problem of large-scale wormholes. In a paper published with Kimyeong Lee of Boston University, he proposed a mechanism for stopping them from forming. "The problem with the quark mass, however, is more serious," said Coleman. "I would say the difficulties fall into two groups: there are those that say that the whole foundations of Euclidean quantum gravity upon which this theory rests are a mess. And they're right. People argue back and forth, and it's very difficult to come to any conclusion, because the subject is so ill-founded. In a way it's like the old quantum theory before the discovery of quantum mechanics, where you had a set of ad hoc rules that were sort of patched together."

"I had hoped when I got into this that the matter would be resolved one way or the other within a short time. I thought the same mechanisms that led to the cosmological constant being zero would lead to a determination of the other constants of nature, and we would therefore see if the theory was right. But unfortunately the only calculation that could be carried out was one for quark mass, and it came out wrong." He pointed out, however, that a very crude approximation had been used in this calculation, and he hopes that further refinements will be made.

The other problem with the theory, and quantum gravity in

general, said Coleman, is that there is little or no contact with experiment. "This subject is in a state where it has neither firm theoretical foundations nor contact with experiment. We need one or the other. If you have a well-defined theory that has been well-formulated and checked in a lot of places, and it predicts something you don't know how to verify, you're still happy. There's black holes, for example, which were predicted in ordinary classical general relativity and were fairly well understood by Oppenheimer and Snyder 40 years before anyone had any experimental evidence for them. That was okay because there was all those other tests of general relativity that had been checked, so you knew that black holes were going to be out there somewhere. This theory doesn't have that."

Coleman says that he is concentrating on other projects now. "I'm not working on wormholes right now," he said, "but one never knows. I could wake up suddenly one morning with a bright new idea. I spent two years on it. But who wants to spend the rest of their life in controversy. You want to do something that is confirmed or rejected, or possibly uncover some new phenomenon in an already well-established theory. If you want to fight all the time, you go into the social sciences [laugh]. My current judgment is that this line of development has not been shut off, but in the current state of our knowledge we have pushed it around as much as it can be pushed."

He concluded the interview with a philosophical reflection. "You go on," he said. "You do physics. You write it up. Life goes on. Eventually you fall over, but you've had a good time."

ANOTHER APPROACH TO WORMHOLES AND OTHER UNIVERSES

Wormholes and other universes also play an important role in some work being done by Alan Guth of MIT and several of his colleagues. Born in New Jersey, Guth received all three of his degrees from MIT, his B.S. and M.S. in 1969, and his Ph.D. in 1972.

Alan Guth.

He has been an instructor at Princeton and a research associate at Columbia, Cornell, and Stanford.

He was not particularly interested in cosmology as a child, but he does recall being intrigued by at least one book he read on the subject either in elementary or high school. He was interested in science, though, and his interests gradually focused. In high school he decided he wanted to study physics, and by the time he had entered college he found that his strongest interest was theoretical particle physics. Nevertheless, he did an undergraduate thesis project in experimental nuclear physics, wanting to have some contact with experiment before immersing himself in theoretical studies. His Ph.D. work concerned the then-novel idea that many "elementary" particles are composed of quarks.

Cosmology was too speculative for him. It wasn't well-formulated enough, he said. But in 1978, after attending a lecture by Robert Dicke of Princeton, he began to take an interest in it. A colleague, Henry Tye, then convinced him to collaborate with him on a problem in the area, and by 1979 things had begun to gel. Within a short time he had published his inflation theory.

The key ingredient of the early universe that led to inflation is what is called a false vacuum, a highly unstable state of high-energy matter. It is from this state that a sudden increase in the expansion rate of the universe occurs, the so-called inflation stage. In fact, all that was needed to produce the entire universe and everything in it 10^{-35} second after the big bang was a tiny region of false vacuum. The total energy content of this tiny region was only about 10 kilograms. If this is all that was needed, one wonders if it might be possible to create a universe from a man-made false vacuum. Guth and colleagues Jemal Guven and Ed Farhi have looked into this possibility. "We've been looking at the question of whether it is possible in principle to create a new universe by assembling a region of false vacuum, and then setting off a localized inflation," said Guth. "First we looked into where this universe would go if it could be created. It turns out it would split off completely from our universe leaving it intact. We then considered whether or not it is possible to create a sphere of false vacuum that is big enough for this to happen. It turned out that if we used pure classical general relativity, it was impossible to do if it did not start from an existing singularity. And a singularity is something we don't know how to make. So we asked ourselves if there was a nonclassical way, a quantum mechanical way, of doing it without an initial singularity."

What they found was that the small sphere of false vacuum could be made to expand, and according to quantum theory it could "quantum tunnel" through nonspace and become a new universe. The mass density that was required to give the required false vacuum, however, was 10^{75} grams per cubic centimeter, far above anything that is technologically possible. It is conceivable, though, Guth says, that a civilization in the distant future might be capable of producing such densities.

Guth and his colleagues found that if someone could, in effect, compress a 10-kilogram sphere of matter into a space approximately one-trillionth that of a subnuclear particle, they could, in theory, produce another universe. This new universe would inflate rapidly from the false vacuum, until its eventual occupants saw it as we see our universe. They noted, however, that although there would initially be a wormhole between this universe and ours, it would soon seal off. It would then be impossible to communicate with it.

I asked Guth if the wormholes he visualized were macroscopic in size. "If it were man-made, you could imagine it being almost any scale," he said. "But if it were to be done with the least possible effort, it would be a universe that starts off with a sphere of false vacuum of microscopic dimensions, approximately 10^{-24} centimeter, which is small, but much bigger than the Planck scale. It would then tunnel through a wormhole of roughly this size."

Guth and his colleagues were particularly pleased when, shortly after they had completed their work, a group at the University of Texas consisting of Willy Fischler, Daniel Morgan, and Joseph Polchinski verified their result using a slightly different method.

When detailed numerical calculations were carried out, however, there was some disappointment. Although the process was, in theory, possible, the probability that it would actually occur was extremely low. Still, Guth is optimistic. "Even with this small probability," he wrote, "there might still be a large probability of an event of this sort occurring somewhere in a universe that has undergone a large amount of inflation." He admits, however, that under the circumstances it would be virtually impossible for man to perform the feat.

So, while there appear to be serious difficulties at the present time, inflation is, nevertheless, a possible method of producing new universes, and of increasing the size of microscopic wormholes.

CHAPTER 13

Advanced Civilizations and Time Travel

We have looked into the possibility of using time tunnels to travel to the future and back to the past. More importantly, though, we would like to know if someday in the distant future these tunnels will be used to travel to the stars. If this does happen, it is reasonable to assume that other civilizations somewhere out there are using them right now. Do we have any evidence that this is the case? Unfortunately, we don't; we can only speculate, but careful scientifically guided speculation is often useful.

EVIDENCE FOR OTHER SOLAR SYSTEMS

Let's begin with planetary systems. If there are civilizations out there, they must have a place to live. So it is logical to begin by asking if there are any stars out in space with planets orbiting them. Our major interest is sunlike stars, since we know that they are the best candidates for sustaining life. But if we can find any evidence for planets around any kind of star, we will know, at least, that solar systems such as ours exist.

The best way to search for evidence of planets is to look for infrared radiation—long wavelength radiation that lies beyond the visible red end of the electromagnetic spectrum. Warm objects here on Earth radiate in the infrared, and the Earth itself is a strong source of infrared radiation. So if there is a planet near a star somewhere out in space, it should emit infrared radiation.

Could we detect the radiation from such a planet? It's unlikely that we could detect it directly; it would be drowned in the radiation from its parent planet. If, for example, we were out in space looking back at the Earth, it would be difficult to distinguish the Earth's radiation from that of the sun. It is best therefore to look for stars that appear to have excess infrared. Stars like our sun, for example, do not radiate strongly in the infrared. If we found such a star, it is quite possible that the excess radiation would be coming from planets associated with it.

One of the first infrared surveys was made by the Infrared Astronomical Satellite (IRAS), which was launched in 1983. As the data began to pour in, a new window was opened; the sky was dotted with infrared sources. Among the most interesting were the stars Vega, Beta Pictoris, and Epsilon Eridani. Brad Smith of the Jet Propulsion Laboratory and Richard Terrile of the University of Arizona took a particular interest in Beta Pictoris. They decided to get a closer look at it with one of the telescopes of the Las Campanas Observatory in Chile. Was the excess infrared coming from an accretion disk, or perhaps a system of planets, in orbit around it? Using a 100-inch telescope they blocked off the light from the star itself and searched the region around it using a supersensitive electronic camera. To their amazement, they found that Beta Pictoris was surrounded by a giant disk of matter. It extended out for 100 billion miles—about 30 times the average distance between Pluto and the sun—and it was made up of hundreds of billions of tiny particles and grains of ice. Yet, strangely, close to the star, space was clear of debris. Astronomers believe that there may be planets in this region. Was this a solar system in formation? Our solar system formed about five billion years ago from a large gaseous disk called the solar nebula. It was out of this solar nebula that the the Earth and other planets were built up through a gradual process of accretion. Tiny grains in the nebula coalesced, forming small rocky "planetesimals," and they in turn struck one another and coalesced until finally there were nine protoplanets circling a protosun in a dense nebular fog. Then, as the temperature of the protosun's core reached 15 million degrees, nuclear reactions were triggered and an explosive wave swept through the solar system clearing out the fog.

Are we seeing the early steps of a process like this on Beta Pictoris? It is possible. Furthermore, we also have evidence for a similar ring around Vega, the bright blue star in the summer sky.

The discovery galvanized the world of astronomy. Astronomers began to look more carefully at the IRAS data. Dana Backman of the National Optical Astronomical Observatory (NOAO) studied the data from approximately 135 stars and found that 25 of them showed excess infrared radiation. Two stars in particular attracted his attention: Gliese 803 and Gliese 879 (from a catalogue of nearby stars by Wilhelm Gliese). Using a 120-inch telescope equipped with an infrared detector, he carefully scanned the two. In both cases there were faint points of light next to the star itself. Were they planets? Backman is still not sure. It is possible that they are background objects, but even if they aren't, they are much larger than Earth, perhaps even larger than Jupiter. They may be brown dwarfs, objects that are in the fuzzy region between stars and planets.

We cannot say for certain that we have detected a planet directly, but with advancing technology astronomers believe that we should be able to within a decade or two. But direct detection and use of infrared are not the only ways of detecting planets. We can also look for them indirectly. If a planet was in orbit around a star, the star would appear to wobble slightly as it moved through space. This is because planets do not actually orbit the center of their star; all of the objects in a system—star and planets—move around their center of mass. In most cases the star is much heavier than the planets, and the center of mass is well inside the star. Nevertheless, the star is still perturbed as it moves through space.

The Canadian astronomers Bruce Campbell, Gordon Walker, and Stephanson Yang used this technique to search for planets around several nearby stars. It is well-known that if an object is moving away from us, its spectral lines are shifted toward the red end of the spectrum because of the Doppler effect. And if it is moving toward us, they are shifted to the blue end. In the case of a star with a slight wobble the shift is small, but Campbell and his colleagues developed a device that is capable of detecting radial velocities (velocities along the line of sight) as small as 10 meters per second. Starting in 1980, they studied 15 stars and found that half of them had a wobble.

Only one of the objects in Campbell's survey, however, has gone through its complete cycle. Called Gamma Cephei, it is an orange subgiant star with a mass slightly greater than that of our sun. Spectral measurements indicate that it undergoes a velocity variation with a period of about 2.6 years. Using this, along with an assumed mass for the star, shows that its companion is about twice as far from its star as the Earth is from the sun, and it has a mass about 1.5 times that of Jupiter.

Using a similar but less sophisticated system, David Latham of the Harvard–Smithsonian Center for Astrophysics and several colleagues have found evidence for a dark object in orbit around the star HD 114762. They have studied it for a total of 12 years. Its orbital period is a mere 84 days, so it is relatively close to its star, probably about the distance Mercury is from the sun. It is, however, quite massive, probably ten times as massive as Jupiter.

The evidence for planetary objects around nearby stars is obviously encouraging. A number of astronomers have suggested that 10 percent or more of stars may have dark companions. But is it likely that any of them contain life?

STATISTICAL EVIDENCE FOR LIFE

Although we have little in the way of direct evidence for extraterrestrial life, we can make an estimate based on probability. Frank Drake of Cornell derived a formula in 1965 that gives such an estimate. It is based on such things as the number of planets per star, the fraction of these planets that are in the ecosphere, or life zone of the star (region around a star where water is liquid most of the time), and the probability that life will form on these planets.

To determine the number of possible good candidates you must first determine what type of star is acceptable. Stars similar to the sun are obviously excellent candidates. Looking into the details we find that if a star is considerably more massive than the sun, it will not live long enough to develop life. We know that life took about 4.5 billion years to develop on Earth, but a star much more massive than the sun, say 15 times as massive, goes through its life cycle in only 15 million years. Small stars are also a problem. Their ecosphere is very narrow and much too close to the star. If a planet stayed in such an ecosphere, it would soon become tide-locked to the star (it would then keep the same face toward the star at all times). Temperatures would be extreme if this were the case: one side would be too hot, and the other too cold for life. Under such circumstances, it's unlikely that life would develop.

The numbers we get from Drake's formula vary considerably, depending on how liberal we are in our estimates. Of particular importance is the fact that they depend critically on the lifetime of a civilization. To a first approximation, we can say that the number of civilizations in our galaxy that we could communicate with now is equal to the longevity of a civilization; let's call it L. As you might expect, though, L is extremely difficult to determine. The only civilization we are familiar with is our own, and L for it is 100 to perhaps 10,000 years depending on exactly how we define longevity (100 years is roughly the number of years we have had technology).

Let's take a closer look at this problem. There are many factors to be considered when trying to arrive at a number that is representative of L. Wars and natural resources are, of course, going to be major factors. Natural resources, particularly energy resources, are a concern because they are limited. Furthermore, they are a function of the population, and we know that the Earth's population is increasing rapidly. In fact, it's doubling every 35 years. Needless to say, it is going to be difficult for us to keep up

With so many stars in the universe the probability of life somewhere among them is high. (Courtesy California Institute of Technology.)

with such a rapid growth. For the same standard of living, everything—food production, housing construction, energy resources, and so on—will also have to double every 35 years.

It's difficult to see how we will be able to accomplish this. Sooner or later we are going to start to run out of things. In 700 years, for example, 20 doublings will have occurred, and for every five people on Earth today there will be over 2.5 million. And everything else on Earth will have to increase at the same rate for us to maintain the same standard of living. It's easy to see that within a few hundred years we're going to be in trouble—unless something drastic happens. It's quite unlikely that there are enough resources left on Earth to support such a population.

Even worse is the fact that our demand for energy is increasing much faster than the general population. The only way we will be able to generate such large energies is if we expand to other planets and start using their resources. Furthermore, we'll eventually have to start using more of the sun's energy. At our present expansion rate we will need all of the energy that the sun generates within a few hundred years.

In addition to this, and perhaps more importantly, as far as longevity is concerned, is the factor of war. How long will a typical civilization last before it annihilates itself in a nuclear war? This is a major concern. If a worldwide war broke out and most of the life on Earth was annihilated, it is unlikely that a civilization coming after us, starting from scratch, would be able to follow our path to advanced technology. Early man on Earth had readily available raw materials, many of them near the surface. We are gobbling up everything that is easily available, so it would be virtually impossible for another civilization starting from a primitive state to advance to a high technology.

So it may be that L is relatively short. If it is, indeed, as low as 1,000 years, or even 10,000 years, there are very few civilizations out in space right now that we could communicate with.

Another thing we have to consider is that the universe is approximately 18 billion years old and it took us only 5 billion years to develop life and a technology. This means that we started rather late. It is reasonable to assume, although there are certainly many factors we cannot be sure of, that most civilizations began much earlier than we did. Many of them would have developed technologies when our galaxy was only 5 or even 10 billion years old. If this was the case, and civilizations last only a few million years, then life was much more common in the past than it is now.

Much of what I have said so far is speculative, but we are certain of one thing: the number of civilizations that are presently in the universe determine how far they are separated on the average. Furthermore, the number of these civilizations depends critically on *L*. Various estimates have been made for *L*; the German astronomer Sebastian Von Hoerner, for example, is convinced that civilizations are relatively rare. His estimate gives an average distance between them of 1,000 light-years. If true, we would certainly have problems communicating with them. It would take 1,000 years to get a radio message to them, and we would have to wait another 1,000 for a reply.

Others, however, are more optimistic. After all, there are 200 billion stars in our galaxy. They estimate that the average distance may be as low as 50 light-years. This is much better, but two-way communication would still be very difficult.

TIPLER'S VIEWS

A number of scientists, however, are quite pessimistic about our chances of detecting another civilization. Some of them go as far as saying that we are the only civilization in our galaxy. Frank Tipler of Tulane University is one of the more vocal of this group. His arguments center around two points: the small probability that an advanced civilization will develop (considering all of the problems), and the high probability that if one did, it would eventually colonize the entire galaxy.

Tipler acknowledges that it would be difficult to get people to take voyages of hundreds of years, which is what would be needed to reach the stars, assuming they didn't use wormholes.

But he is convinced that any advanced civilization would send out computer-controlled probes, similar to the ones we sent to the moon and Mars, though much more sophisticated. At 10 percent the speed of light, such a probe would get to Alpha Centauri, our nearest star, in about 45 years.

According to Tipler, when a probe got to the planet of a nearby star, it could use its resources to build several clones of itself using preprogrammed computer instructions. Each of these clones could then be sent to another star. Allowing, say, 100 years for travel and, say, 500 for establishing a base and building clones, it would take only a few million years for these probes to move across the entire galaxy. But our galaxy is approximately 16 billion years old; a few million years is only a tiny faction of its life. If this has, indeed, happened, there should be probes everywhere. Since we have not seen any, Tipler is convinced that we are the only advanced form of life in our galaxy.

Michael Papagiannis of the University of Boston gives a similar argument, but he considers a different mode of travel. He assumes that humans would make the trip in huge spacearks that could accommodate hundreds of people. And again, he allows 500 years for a trip of 10 light-years and 500 years for establishing a colony, and finds that the entire galaxy would be populated in only 10 million years. Since there are no signs of them, life must be sparse. Like Tipler, Papagiannis believes that we may be the only advanced life in our galaxy.

You could, in fact, extend this line of reasoning even further. If advanced civilizations have developed time tunnels, it would be even easier for them to travel from star to star, and there should therefore be some sign of them.

As you might expect, there is strong opposition to this point of view. Frank Drake of Cornell argues that colonization would take much longer than Tipler and Papagiannis have suggested, mostly because of energy problems. Jill Tarter of the Space Science Laboratory at NASA has pointed out that a roundtrip starship flight of 10 light-years with a payload of 1,000 tons and at a speed of 70 percent that of light would require an energy expenditure

500,000 times that used in the United States in an entire year. This isn't to be taken lightly.

Would a civilization want to undertake an endeavor with such a large energy expenditure? The question of motivation is important. All in all it's a difficult and controversial issue. And it's hard to say that just because we haven't seen anyone yet, there's no one there. Furthermore, Carl Sagan argues that a random diffusion of the sort described by Tipler would take roughly the age of our galaxy to completely populate it.

At any rate, the only way to really decide the issue is to continue searching for evidence that they do, indeed, exist, or perhaps prove that they don't.

LARGE-SCALE RADIO SEARCHES

The only way for us to find out for sure if there are any advanced civilizations out there is to search for them. And the best way to do this is with radio telescopes.

Unfortunately, there's a problem: the universe is noisy, and a signal from another civilization would easily be drowned out. Furthermore, it might be absorbed in our atmosphere. Is there any way we can avoid these problems? Indeed, there is, at least to some degree. First, we know that the amount of noise in space varies with frequency. In other words, the electromagnetic spectrum is noisier at some frequencies than others. Second, we know that absorption of a signal also varies with frequency. We can therefore avoid the frequencies that are problematic.

At low frequencies stars are noisy, and at high frequencies the noise from our atmosphere interferes with the signal; both of these regions should be avoided. Between them, however, is a range of frequencies that is relatively quiet; it extends from 1 to 10 GHz (a GHz is a billion cycles per second). In 1959, Guiseppe Cocconi and Philip Morrison, who were then at Cornell, suggested that the region near 1.4 GHz was our best bet. This is, interestingly, where hydrogen atoms broadcast a radio signal to space; furthermore, it's just below where the radical OH broadcasts. Since hydrogen and OH together make water, it is sometimes referred to as the "water hole."

Having decided that this is the best region to search, we are then faced with the problem of what to search for. Is it likely that a civilization out there is beaming a message to us? We have to admit that this is quite unlikely. Nevertheless, if there was a civilization with radio antennae, say, 20 light-years away from us, they could detect us. Beginning about 50 years ago, UHF-TV stations began broadcasting, and much of their broadcast radiation escapes to space. This means that we are surrounded by an expanding sphere of "stray radiation" that extends out about 50 light-years. Of course, the sun also gives off considerable radiation, but an advanced civilization would easily be able to distinguish it from Earth's. Interestingly, in some regions of the spectrum, the Earth would be an even brighter source than the sun. Once this civilization distinguished the two sources, they could learn a considerable amount about us: the length of our day and our year, our distance from the sun, and our average temperature.

Our sphere of radiation is still moving outward. In another 50 years it will extend out to 100 light-years, enclosing a volume that includes roughly 1,000 stars. If there were an advanced civilization near any of these stars, they would be able to detect us.

In the same way that civilizations out there could detect us, we could detect them. And over the years many projects have been set up in hopes of doing just this. In 1972, for example, Zuckerman and Palmer of UCLA listened to 600 stars over a period of four years. They observed fluctuations in some of the signals, but nothing indicating a civilization. An even larger survey was initiated by Ohio University in 1977, but they also discovered nothing of importance. One of the most extensive of recent surveys was one initiated by Paul Horowitz of Harvard in 1983. He had an advantage over his earlier counterparts; with his equipment he was able to listen to thousands of different frequencies at once.

One of the major problems in a search such as this is the

number of different frequencies that have to be listened to. It may seem that the range 1 to 10 GHz is small, but if you listened carefully to each 1-hertz wide channel of this region for even a few seconds, it would take several thousand years to complete the survey. It is therefore advantageous to be able to scan many channels at once.

Horowitz was able to do this. With his original system he was able to listen simultaneously to 131,000 channels in the vicinity of 1.4 GHz. In 1985 he increased this to 8 million channels. So far, however, he has not detected anything that looks like an artificial signal.

An even more ambitious program will be launched during the 1990s by NASA. Over a period of ten years, all frequencies from 1 to 10 GHz will be listened to using 8 million channels. In this project, 800 of the best candidates, all similar to the sun, and all within 100 light-years of it, will be listened to. With such a program, it is reasonable to assume that if there is, indeed, a technological civilization within 100 light-years of us, we would detect it.

SUPERCIVILIZATIONS

A number of astronomers have suggested that we might be better off searching for civilizations that are considerably more advanced than us. This brings up the question of how advanced a civilization can become. This is important as far as wormholes are concerned, since it is obviously going to take a very advanced civilization to perfect and use them. I'll have to admit, though, that when we start talking about things such as this, we can do little more than speculate.

The Soviet astrophysicist N. S. Kardashev was one of the first to consider this problem. He divided civilizations into three groups according to their energy use, referring to them as Types I, II, and III. Type-I civilizations are those that control the energy resources of their planet. In particular, they have controlled thermonuclear fusion. We are almost at this stage.

Type-II civilizations are those that use most of the energy of their star. Interestingly, at our present rate of development we will need all of the energy emitted by our sun in about 500 years if we are to keep a reasonable standard of living. In the late 1950s, Freeman Dyson of the Institute for Advanced Study in Princeton suggested a way that this might be possible. We could, for example, put a swarm of small satellites in orbit around the sun, each a collector. Several hundred thousand, however, would be needed to collect all or most of its energy. Such a sphere is now referred to as a Dyson sphere.

For us, stage II may not be far off if we continue to use energy as we are now. On the basis of this, it seems that there should be quite a few Type-II civilizations in our galaxy right now. Dyson has shown that it would only take about 2,500 years to go from Type I to Type II.

Beyond Type II we have Type III. Kardashev defined them to be civilizations that control the resources of their entire galaxy, or at least a large fraction of it. In principle, they should be easier to detect than Type-II civilizations, but in practice it might be more difficult in that we would have to look in other galaxies. Furthermore, if we did find one, or evidence for one, we certainly wouldn't be able to communicate with it because of its distance.

Getting back to Type-II civilizations we ask: How would we go about looking for them? Dyson has suggested that the best way would be to look for infrared sources. But as we saw earlier we have spent a considerable amount of time in recent years looking for infrared sources. Most of the sources that have been found so far, though, are dim red stars. No evidence for Dyson spheres has been found. But as Dyson points out, "We've barely begun to search."

In searching for Type-III civilizations we would search for a large number of infrared sources throughout a galaxy. The overall galaxy would emit most of its radiation in the infrared, and would therefore be relatively easy to spot. It might be possible, though, that as the civilization spread through the galaxy, it selected only certain types of stars to use as energy sources. If this was the case, the civilization would be much more difficult to detect.

A Type-III civilization would control its entire galaxy and would, perhaps, be able to travel to neighboring galaxies. Several galaxies are shown in this photograph.(Courtesy National Optical Astronomy Observatories.)

Dyson is convinced that the best way to find a Type-II or Type-III civilization is to look at stars and galaxies in as many ways as possible. Study all of the electromagnetic radiations emanating from them: infrared, ultraviolet, X ray, and so on. In short, become as familiar as possible with natural sources so that you can easily distinguish an artificial one.

COULD AN ADVANCED CIVILIZATION BUILD AND STABILIZE A WORMHOLE?

If there really are other civilizations in the universe, and some of them are much more advanced than us, we ask: Could they build a stabilized wormhole? Or perhaps we should begin by asking: In order to be able to build a wormhole, at what stage of development would a civilization have to be in? We are on the verge of becoming a Type-I civilization, and assuming that we survive, we will move on to being Type II in about 500 years. It is quite likely that we will become a Type-II civilization before we have the technology for making and stabilizing wormholes. Type-II civilizations are, in general, still restricted to their own star, but are in the process of developing techniques that will enable them to travel to other stars. In fact, they cannot advance to Type III until they have developed these techniques. Because of the tremendous distances involved, it seems that control of wormholes, assuming that it happens, will come only when a Type-II civilization is quite advanced.

What exactly would constructing a traversable wormhole entail? You would obviously have to start by constructing an unstable one. As we saw earlier, it has been suggested that you could pull one out of the quantum foam, but, of course, we have no idea how this would be done. Another possibility is through inflation from a false vacuum. But again the problems are severe.

Once we had a wormhole, there's the problem of stabilizing it with exotic matter. We don't know what form this exotic matter would take or how we would handle it. Again, these are engineering problems. We do have some idea how to make them into time tunnels using strong gravitational fields or by moving them rapidly. So we could, in theory, use them to go to the past or the future. But there's also the problem of how we could use them to travel to nearby stars. In Sagan's novel *Contact*, they were used to travel to Vega and the center of our galaxy in a matter of minutes. Something like this will no doubt remain in science fiction for many years to come, unless, as in *Contact*, we stumbled on one made by another civilization.

NETWORKS OF WORMHOLES

Certainly if a civilization did manage to use wormholes to get to some of its nearby stars, within a relatively short time there would be a whole network of wormholes set up all around it. A Type-III civilization would no doubt have a network of this type, just as we have networks of subways under many of the major cities of America.

The problems in setting up something like this are hard to imagine, but if it is possible, a Type-III civilization would almost certainly use it. Is there any way we could detect such a network? Unless there were an opening close to Earth, this is quite unlikely. It is not a system you could see with a telescope, it's merely curved space.

If we did find firm evidence of a Type-III civilization in a nearby galaxy, it would, perhaps, tell us that a network of wormholes is possible. Communication and travel via ordinary routes over a large fraction of a galaxy would be so difficult that it's virtually impossible that a Type-III civilization could be under central control without such a system. Can you imagine sending a message from an outpost back to home base via a radio telescope and waiting for a reply. It could take 200,000 years.

We could go even further and imagine a Type-IV civilization. This would be one that controlled the energy sources of a cluster of galaxies. If there were such a civilization, it would certainly have developed wormhole technology.

But what if there are problems we haven't foreseen and wormhole technology isn't possible. Is there any other way to travel to the stars?

CHAPTER 14

Faster Than Light?

The major problem with travel to distant parts of the universe is that we are restricted to speeds less than that of light. We can get around this by using time tunnels, assuming that they are possible. For now let's ignore time tunnels and consider other possibilities. We know that the time dilation of special relativity allows us to travel at "effective speeds" well in excess of the speed of light. The reason for this is that an observer on Earth notices that time passes very slowly on a clock in a spaceship that is traveling at a speed very close to that of light relative to the Earth. The astronauts in the spaceship don't notice this slowing; to them time appears to pass normally. But if we use the slowed-down time seen by Earth observers to calculate the "effective speed" of the spaceship, we obtain numbers well in excess of the speed of light (usually called c).

For example, if we accelerated with an acceleration equal to Earth's gravity (one g) for five years, we would reach an "effective speed" about 75 times that of light ($75c$). Continuing at the same acceleration we would, in a matter of a few years, be traveling at an effective speed thousands of times that of light. Our real speed—in other words, the speed we measure—would, of course, always be less than that of light. Nevertheless, as far as travel to distant points of the universe is concerned, it's the effective speed that counts. Our only problem is that we would have to forget about any ties to our home planet. Thousands, or even millions of years would pass back there during our trip, and if we returned, no one would even remember that we had left.

But is it possible to travel at a "real" speed greater than c? In this case we would be able to visit distant parts of the universe and come back to Earth in a relatively short time, say, a few months after we left. This is, of course, done routinely in science fiction. Astronauts jump into their spaceships, accelerate to near c, then use hyperdrive to jump to speeds greater than that of light. We know, of course, that this can't be done with our present technology. But is the region beyond c completely forbidden to us? Let's look into this.

TACHYONS

One of the first to consider what it would be like to travel at speeds in excess of c was the German physicist Arnold Sommerfeld. About 1900, he used Maxwell electromagnetic equations to show that particles with speeds greater than c would speed up when they lost energy. This seemed strange as it was well-known that particles with speeds less than c slowed down when they lost energy.

A few years later, however, Einstein came along and told us that we needn't worry about such things. The speed of light was unattainable, and therefore speeds beyond it couldn't be reached. After all, to get to them we would have to pass through c.

But why is c unattainable? According to Einstein, as a particle's speed increases it becomes heavier. It therefore takes more energy to move it. In fact, it takes an infinite amount of energy to take it all the way up to the speed of light. Since the universe doesn't contain this much energy, the speed of light is unattainable.

In the early 1960s, however, a number of physicists including Gerald Feinberg of Columbia University, George Sudarshan, now of the University of Texas, and Olexa-Myron Bilaniuk of the Ukraine, began taking a second look at Einstein's equations. It was easy enough to see that c was unattainable, but if you could somehow overcome c, in other words, jump it to greater speeds, Einstein's equations were well-behaved. The only difference was that imaginary quantities now appeared, but this didn't worry them.

In a sense, there was another world beyond the speed of light, a world where particles traveled only at speeds greater than c. Feinberg called these particles tachyons. But there was still the problem of crossing the barrier at c. Was there a way around it? Indeed, there was: if the particles were born with speeds greater than c, and never traveled at c or less, you didn't need to worry about the barrier. There were, however, other problems. According to Einstein's equations, if these particles existed, they would have an imaginary mass. In other words, if you squared the mass you would get a negative number. None of the particles in our world have this property, so it was difficult to understand what it meant.

Furthermore, there was an even more serious obstacle. We know that when we observe a clock moving at high speed relative to us, it appears to run slow compared to our clock. If it could travel at c, in fact, it would stop. This means that if a particle somehow passed through c, its clock would run backward compared to ours. And this, in turn, creates a causality, or time-ordering problem. If a tachyon was created, then collided with another particle, would the collision occur before it was created, or after? We're not sure. But if it came before we would obviously have a serious problem.

PROPERTIES OF TACHYONS

We have seen that tachyons would have an imaginary mass, but they would also have other strange properties. As they lost energy, they would speed up; eventually, with very little energy, they would be traveling millions of times the speed of light. Indeed, when they lost all their energy they would travel at an infinite speed. Tachyons with this property are called "transcendent" tachyons. It is, in a sense, their rest state.

If a tachyon is in a transcendent state, you can slow it down by supplying energy to it. But it would take an infinite amount of energy to take it from this state all the way down to the speed of light. This is, of course, impossible, so we say that c is unattainable (in a downward direction) for tachyons.

Let's come back to their imaginary mass. More specifically, it's their theoretical rest mass that is imaginary. How do we get around this? The adherents of tachyons (and there are very few of them) argue that this is not significant. Tachyons, they point out, are never at rest, so the concept is not meaningful. More important are momentum and energy, and tachyons can, indeed, have positive momentum and energy.

The world, according to tachyons theorists, is made up of two types of particles: tachyons and tardyons. The tachyons travel only at speeds greater than c; the tardyons travel only at speeds less than c. Actually, there's a third type of particle, the luxons, that travel only at c. The photon is the major particle of this family, but gravitons and, possibly, neutrinos also belong to it. There is a barrier at c between these two worlds, a barrier that can never be crossed in either direction. Does this mean that tachyons are forever cut off from us? Let's consider this.

DETECTING TACHYONS

How would we go about trying to detect tachyons? Obviously we would have to search for something that "signaled" their presence. We know that no particle can travel at a speed greater than that of light in vacuum. But if a light beam passes through a medium such as water, it travels slower than the speed of light in vacuum. Because of this, it is possible for particles traveling in water to travel faster than a light beam travels in water. But when this happens, the particles give off a slight bluish glow. In other words, they radiate. This radiation was discovered by Pavel Cherenkov of the USSR in 1934. He wasn't sure what it was when he discovered it, but it was explained a couple of years later by Ilya Frank and Igor Tamm of the USSR. The three men received the Nobel prize for their discovery in 1958.

If tachyons travel at speeds greater than c, they should also give off Cherenkov radiation. The way to detect them, then, would be to look for this radiation. And several experiments have been set up using vacuum chambers surrounded by thick lead shields. If tachyons are charged, we would expect to see a flash of light across the chamber. Furthermore, if we could arrange for the tachyons to collide with other particles, they would lose energy and speed up. Their excess energy would be radiated off as light and we would see a flash in the chamber. So far, though, no unaccounted-for flashes have been seen in any of the experiments.

Another place we might detect tachyons is in cosmic rays. Energetic rays from space, called primary cosmic rays, strike the molecules of our upper atmosphere creating particles that cascade down through the atmosphere. These cascading particles are called secondary cosmic rays, and like the primary ones they travel at speeds close to that of light. The secondary particles also collide with the particles of our atmosphere as they move downward, creating "showers" of new particles. What, you might ask, would happen if a tachyon were among these cosmic rays? Since it travels faster than c, it would reach the Earth before the shower. In fact, since showers take about 20 microseconds to reach the Earth, it would be appropriate to look for events that preceded showers by about 20 microseconds. In 1973, two Australian physicists, Roger Clay and Phillip Couch of the University of Adelaide, reported that they had detected events in a cosmic ray detector just prior to the appearance of showers. They believed they might be tachyons. So far, however, they have not been able to repeat the experiment, and no one else has verified it.

DO TACHYONS EXIST?

Most physicists are skeptical of tachyons, and the reason is obvious. We have found no evidence whatsoever that they exist.

George Sudarshan of the University of Texas is convinced, however, that we just haven't searched hard enough. He says that most searches have been halfhearted. Feinberg, on the other hand, is satisfied with the searches that have been made. He points out that over 200 papers have been published on tachyons, and although at one time he hoped that they would be found, he now has little faith that they will be.

But it isn't only the lack of experimental evidence that hampers the idea. The foundations of the theory are also shaky. There are the problems of imaginary rest mass and causality violation, but there's also a problem in relation to quantum theory. It is possible to set up a quantum theory of tachyons. Just as we use a superposition of waves—a wave packet—to represent particles in ordinary quantum mechanics, we can obtain a wave packet that represents a tachyon. But when we calculate the speed of this wave packet, it is always less than c. So we have the particle, the tachyon, traveling at speeds greater than c, and the wave packet that represents it traveling at speeds less than c. This never happens with ordinary quantum mechanics and is a sure sign that the theory is flawed.

Bilaniuk and Sudarshan have attempted to put the theory on a firmer basis. They have shown, for example, that under certain circumstances the energy of the tachyon is negative. But a negative-energy particle traveling backward in time can be interpreted as a positive-energy particle traveling forward in time. Others soon pointed out, however, that despite this interpretation, tachyons still violated causality.

The case for tachyons is obviously grim. Most physicists now dismiss them as a bad idea, and few take them seriously. So they may be just an artifact of the mathematics. Interestingly, though, they crop up in modern theories that physicists do take seriously. String theorists, for example, are continually bumping into them in the mathematics of string theory. After cursing a little, they get rid of them by adjusting the theory.

But if tachyons really do exist, could they be of any use to us? Earlier I talked about spaceships shifting to hyperdrive and jumping to speeds greater than c. If this, indeed, happened, the ship would have to undergo a sudden change in construction: from tardyons to tachyons. Is such a change possible? Certainly no one knows, but it seems unlikely. Also, there remains the problem of how we would bridge the gap at c. We couldn't pass through it; somehow we'd have to jump it, and we don't know how to do this.

OTHER METHODS OF OBTAINING SUPERLUMINAL SPEEDS

Are there any other methods of obtaining superluminal speeds (speeds greater than c)? Interestingly, we do occasionally observe such speeds here on Earth. Waves, it turns out, have two speeds associated with them: one is called phase speed and the other is called group speed. To understand the difference, it is best to begin with the simplest wave, what is usually called a sine wave. It has a specific frequency (the number of humps that pass you per second) and wavelength (distance between equivalent points of the wave). Its speed is referred to as the phase speed. This speed depends on the medium the wave passes through, but it is possible for it to be greater than c. Such speeds are observed experimentally in plasmas.

If we superimpose several of these sine waves, we obtain a more complex wave. Joseph Fourier, in fact, showed many years ago that any wave, no matter how complex, can be made up by superimposing sine waves. The speed of this complex wave, or group wave as it is usually called, can never be greater than c.

But if the phase speed of a wave can be greater than c, is there any chance we could "hop aboard" one and travel at speeds greater than c? Actually, it is more realistic to ask if we could send a message via a phase wave at speeds greater than c? To get a message into a wave, however, we would have to distort the wave in some way. But the moment we distorted it, it would become a group wave, and group waves can travel only at speeds less than c. We therefore have a catch-22, and as a result messages can't be sent at speeds greater than c.

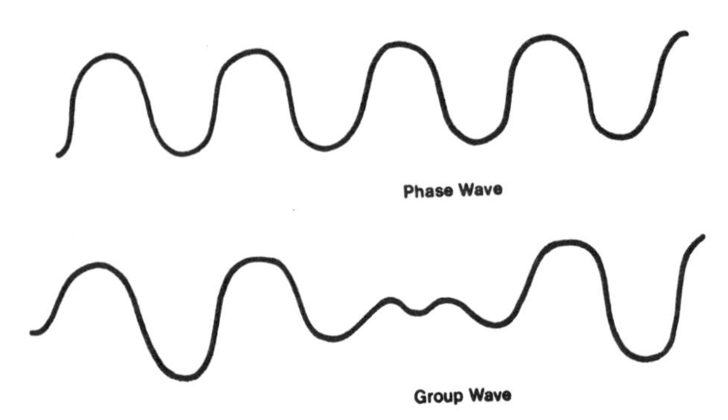

A phase wave and a group wave.

Another interesting time-travel possibility centers around what are called antiparticles. Corresponding to every known type of elementary particle in the universe there is an antiparticle. For example, corresponding to the electron there is the positron, and corresponding to the proton there is the antiproton. Particles and antiparticles annihilate one another when they come together, with the release of one or more photons.

Richard Feynman of Caltech has shown, however, that as far as the mathematics is concerned, there is no difference between an antiparticle traveling backward in time and a particle traveling forward in time. In other words, positrons traveling backward in time act exactly like electrons traveling forward in time. Unfortunately, in laboratory experiments we see no indication of this reversal of time. Electrons and positrons look like ordinary particles except when they collide. There is no physical evidence that time is actually traveling backward for the positron.

But if time did, indeed, flow backward for antiparticles, could we somehow use it to go into the past? I'm sure no one would put any money on it.

THE QUANTUM CONNECTION

In devising quantum theory, Schrödinger showed that when two particles interact, then move apart, the quantum "wave packet" that represents them never completely separates. There is, in effect, always a link between the particles. Over the years there was considerable controversy surrounding this mysterious link. It seemed to be instantaneous, but if this was the case, the two particles could communicate with one another at a speed greater than c. Most physicists refused to believe this.

The problem came to a head in 1935 when Einstein, along with colleagues Boris Podolsky and Nathan Rosen, published a paper in *Physical Review*. Einstein had, for years, been strongly against the ideas of quantum mechanics; in particular, he disliked the aspect of uncertainty. The uncertainty principle tells us that we can't measure two variables such as position and momentum (mass times velocity) simultaneously to high accuracy. Einstein was sure we could.

In his paper, Einstein envisioned two particles that came together, interacted, then separated. One of the particles remained nearby and the other moved off into space. He pointed out that we can easily measure the momentum of the nearby particle, and that would enable us to get the momentum of the distant particle by using the conservation of momentum. (This principle tells us that the total momentum of any system always remains the same.) In making a measurement of momentum, we disturb the system so we can't accurately measure the position of the particle. But if someone out in space simultaneously measured the position of the distant particle, both its position and momentum would have been measured to a high accuracy at the same time. We have, in effect, circumvented the uncertainty principle.

This eventually became known as the Einstein–Podolsky–Rosen paradox (or EPR paradox), and for years scientists argued about it. Then, in 1964, John Bell of CERN published a paper showing that a test could be made to see if the paradox was, indeed, valid, and whether the uncertainty principle could be circumvented. But it was many years before the technology for carrying out the test was available. Finally, in 1982 the French physicist Alain Aspect and several colleagues at the University of Paris were able to perform the experiment. The result was surprising. Einstein was wrong; the uncertainty principle couldn't be bypassed. The test, however, told us more than this. It indicated that there definitely was a quantum link between the systems; they were mysteriously correlated. Furthermore, the link had to be instantaneous.

Was it possible that the systems could actually communicate at speeds in excess of c? Most scientists are sure that this isn't possible. In fact, in 1978 the Berkeley physicist Philippe Eberhard showed that even if superluminal effects occurred in quantum mechanics, we would never see them in the final result. What we calculate in quantum mechanics are average values over many events. The superluminal effects can never affect these averages; they can only occur at the individual event level. In effect, an individual "quantum jump" could occur at a speed greater than that of light, but nothing that we could actually measure.

Could we ever take advantage of this to send messages at a speed greater than c? Most physicists would say it is unlikely. Nick Herbert of Stanford, however, believes that it might be possible, mostly because physicists are still generally quite ignorant of how individual quantum events take place. He writes, ". . . tapping into the network of faster-than-light connections that link up distant quantum jumps all across the universe seems to me to be a wide-open possibility," in his book *Faster Than Light*. So while most physicists are skeptical about the possibility of faster-than-light communication, a few are still hopeful.

CHAPTER 15

Epilogue

This concludes our brief look into the mysteries and enigmas of time travel. We have seen that space and time are intertwined and have properties that only a few years ago were unimaginable. Yet, strangely, we are still not certain that time travel is, in fact, possible. We do know, though, that time tunnels—regions where space and time are twisted into wormholelike tunnels—are possible. Black holes have such tunnels associated with them. But unfortunately they are not traversable. If you tried to pass through one, you would be stretched and crushed beyond recognition. Black holes are, nevertheless, intriguing objects, and one day we may explore them using probes.

Because of the obstacles, for many years there was little hope that time travel would ever become a reality. Then, in the late 1980s, Kip Thorne and his students began studying wormholes that were not associated with black holes. And—lo and behold—they found that they could be made traversable.

But could we construct such a wormhole? Could we actually use it for time travel? Several groups are now looking into this. Without a doubt, many uncertainties remain, a major one being the causality principle. Can it be violated? We don't know. Another is the theory that predicts wormholes: the general theory of relativity. It is a time-tested, reliable theory, but it is not the ultimate theory. We still do not have a quantum version of it.

Considerable effort has gone into trying to construct a quantum theory of gravity, but little progress has been made. The problems have been so severe, in fact, that most of the research in recent years has been directed toward a more general theory, a unified theory, or theory of everything, that includes quantum gravity. This theory would unify all of nature; in particular, it would explain all of the elementary particles and forces of nature.

Scientists have approached this unification on many fronts. One of the most promising theories at the present time is superstring theory. It may be years, however, before it is molded into a completely successful theory, if indeed it ever happens. But even if it doesn't, scientists will continue searching. Furthermore, they realize that if they are to succeed, further breakthroughs are needed. And to get these breakthroughs they will need innovative new ideas. Some of these ideas will, no doubt, seem crazy at first. Relativity theory and quantum mechanics were both considered outlandish when first published. But they are now well-established theories.

If time travel is to become a reality, we still have much to learn. Few, if any, would be brave enough at the present time to say that time travel is just around the corner. Indeed, most of the current work is not a direct attempt to make it a reality, but rather an attempt to see whether it is possible in principle. And we know that even if it turns out to be possible, the engineering problems are so great that it will likely be hundreds, or perhaps thousands, of years before it is finally realized.

So far, interestingly, there appears to be nothing in physics that prevents time travel. Furthermore, we know how it might be accomplished. But actually achieving it is something else. The difficulties are formidable. Where do we get wormholes? Could we stabilize one if we got one?

In conclusion it is perhaps appropriate to briefly outline the major problems.

1. A better understanding of space and time is needed, particularly at the Planck scale.
2. A quantum theory of gravity, or a more encompassing theory that includes it, is necessary.

3. A method for producing wormholes is needed. The quantum foam is one possibility; inflation from a false vacuum is another.
4. A determination of what exotic matter is best to stabilize a wormhole, and an understanding of how it could be applied, is important.
5. And finally, a search for other civilizations is needed. If they were found, and appeared capable of traveling to the stars around them, we would have hope that such travel is possible.

Glossary

Absolute temperature Temperature scale with zero degrees as lowest possible temperature.
Acceleration The rate at which the speed of an object changes.
Accretion disk Flattened disk of matter around a star or black hole.
Aether A hypothetical substance believed at one time to permeate the universe. Needed to propagate waves.
Angular deflection Deflection of a particle or beam of light from its normal path. Measured in degrees.
Angular momentum A measure of the spin of an object.
Anomaly An irregularity.
Antigravity Repulses rather than attracts objects.
Antiparticles Corresponding to every type of particle there is an antiparticle. When a particle and an antiparticle meet they annihilate one another with the release of energy.
Asteroid Rocky bodies that range from a few hundred miles in diameter down to a fraction of an inch.
Axion Particle that is predicted by grand unified theory. Very light.
Baby universe Small universe distinct from ours.
Big bang Explosion that created our universe.
Binary star system Double star system.
Black hole A region of space-time from which nothing, not even light, can escape.
Capillary action The rising of a liquid in a small diameter tube.

Causality The principle that says that cause must come before effect.
Centrifugal force Outward force that is present when an object rotates.
Cherenkov radiation Radiation emitted by an object that is traveling faster than the speed of light in a medium.
Classical theory Any nonquantum theory. General relativity is a classical theory.
Cosmological constant A constant that Einstein introduced into the equations of general relativity to keep the universe stable.
Conservation of energy Principle that states that energy must be conserved, or remain the same, in any physical process.
Constellation A group of stars.
Cosmic ray Highly energetic particles from space that strike our atmosphere creating other particles and radiation.
Cosmology Study of the overall structure of the universe.
Covariance Implies that the form of the equations remains the same in any transformation.
Curved space A distortion or twisting of space predicted by general relativity. Equivalent to gravity.
Degenerate pressure Pressure due to degenerate matter. Occurs in white dwarfs.
Density Mass per unit volume.
Differential geometry Basic geometry of general relativity.
Doppler effect A change in wavelength of waves emitted by a body that is either approaching or receding.
Dyson sphere A hypothetical sphere that is placed around a star to collect its energy.
Ecosphere Region around a star where water is liquid most of the time. Life zone.
Einstein–Rosen bridge A space-time tunnel that has a black hole at its center. Bridges two different universes or two different points of our universe.
Electromagnetic force The force that arises between charged particles.

Electromagnetic wave Wave that is given off by oscillating electrical charges.
Electron Fundamental particle of nature. Negative charge.
Ellipse An egg-shaped curve. The planets travel in elliptical orbits.
Ellipsoid Solid figure with shape of egg.
Emission line A bright spectral line.
Entropy A measure of the disorder of a system.
Equipotential A surface where all points along the surface have the same potential.
Ergosphere Region between the event horizon and static limit of a black hole. Energy can be extracted from this region.
Event horizon Surface of a black hole. A one-way surface.
Exotic matter Matter with tension that exceeds its mass-energy. Matter with negative energy density.
False vacuum An energy state of the early universe. Not a true vacuum.
Force A push or pull on an object.
Frame of reference A system that is used as a standard, to which everything can be compared.
Frequency Number of vibrations per second.
Fundamental constants Constants of the universe such as speed of light and Planck's constant.
Galaxy A large system, or island universe, of stars.
Geodesic Shortest (or longest) distance between two points.
Globular cluster A cluster of stars. Ranges from about 3 million stars down to less than 100,000.
Group speed Speed of group waves.
Hyperbola One of the open conic curves.
Hyperdrive A hypothetical system that allows one to travel at speeds greater than that of light.
Hyperon A type of elementary particle.
Hyperspace A higher dimensional space. Beyond our three dimensions of space.
Imaginary mass The square root of the negative of mass.

Inertia Resistance to a change in motion.

Inflation theory A theory that suggests that a sudden increase in the expansion rate of the universe occurred shortly after the big bang.

Infrared radiation Radiation that has a wavelength slightly longer than red light.

Inner event horizon All black holes except Schwarzschild have two event horizons. This is the one closest to the singularity.

Irreducible mass When black holes lose spin or charge they decrease in mass. This is lowest possible mass.

Interstellar travel Travel between stars.

Kerr black holes A spinning black hole.

Kruskal coordinates A coordinate system for the region around a black hole.

Lagrangian point Point between two bodies where gravitational potential is equal.

Light cone In space-time diagram, speed of light is represented by a line at 45 degrees. If this diagram is spun around its time axis, you get a light cone.

Magnetic field line Line of force around a magnetic field.

Manhattan Project Name of wartime project in which atomic bomb was built.

Mass A measure of the quantity of matter in a body.

Meson A medium-weight elementary particle.

Naked singularity Singularity that is not enclosed in an event horizon.

Negative curvature A type of curvature. A saddlelike curve has negative curvature.

Nuclear reaction An energetic reaction involving changes in nuclei.

Nucleon Particle such as proton or neutron.

Neutrino An elementary particle. May have no mass. Extremely elusive.

Neutron star A star that is made up mostly of neutrons. Usually only a few miles across.

Newton's theory Theory of motion devised by Newton.
Parallax The apparent shifting of a nearby star relative to distant background stars as the Earth moves around in its orbit.
Parsec A measure of distance equal to 3.26 light-years.
Penrose diagram A space-time diagram devised by Penrose. Infinities are shown in diagram.
Phase speed Speed of phase waves.
Photoelectric effect The emission of electrons from a metal when light is shone on it.
Photon A particle of light.
Photon sphere Surface around a black hole. Lies 1.5 times farther out than event horizon.
Pinch off Refers to pinching off of the throat of a black hole or space-time tunnel.
Pion A medium-weight elementary particle.
Planetesimal Small rocks in the early solar system. Coalesced to produce protoplanets.
Planck time 10^{-43} second.
Planck length 10^{-33} centimeter.
Positive curvature A type of curvature. The surface of a ball has positive curvature.
Positron Antiparticle of electron. Has positive charge.
Polygon A many-sided object.
Precession A change in position of the major axis of an ellipse.
Primordial black hole Black hole created in the big bang.
Protoplanet Early form of a planet.
Pulsar A very-short-period variable star. Composed of neutrons.
Quantum coherence A measure of coherence or consistency in a quantum system.
Quantum field theory Quantum theory of a field (e.g., electromagnetic field).
Quantum gravity A quantized theory of gravity.
Quantum jump A jump that an elementary particle such as an electron takes between energy levels.

Quantum theory Theory of atoms and molecules, their structure, and interactions with radiation.

Quantum wave function A mathematical function that describes a quantum system.

Quark An elementary particle. Protons, neutrons, and most other heavy particles are made up of quarks.

Quasar Energetic object in the outer regions of the universe.

Radiation A form of energy. Photon.

Radio galaxy An energetic galaxy. Gives off radio waves.

Radial velocity Speed in a direction directly away or toward us.

Radio telescope Telescope used for detecting radio waves.

Redshift A shift of spectral lines toward the red end of the spectrum. Caused by the Doppler effect.

Ring singularity Singularity in Kerr black hole in the form of a ring.

Roche lobe Gravitational equipotential that has the form of a figure eight around two gravitating bodies.

Schwarzschild radius Radius at which escape velocity is equal to the velocity of light.

Singularity A region where a theory such as general relativity goes awry and gives incorrect answers. Point or ring at the center of a black hole.

Space-time A four-dimensional unification of space and time.

Space-time diagram A diagram in which space is plotted along the horizontal axis and time along the vertical axis.

Spectral line Line obtained when light from a star or other object is passed through a spectroscope.

Spectroscopy The study of spectral lines.

Special relativity Theory that applies only to uniform, straight-line motion.

Starquake A cracking that occurs in the crust of a neutron star.

Static limit Surface around a black hole, inside of which nothing can remain static.

Statistical mechanics The study of atomic and molecular systems using statistics and probability.

Stellar evolution The changes that take place in a star as it ages.

Stray radiation Refers to radiation from TV stations and so on that escapes from the Earth.
String theory A theory that assumes elementary particles are made up of tiny strings.
Supergiant star Large red star. If one were at the position of the sun, it would extend out past the orbit of Earth.
Superluminal Faster than the speed of light.
Supernova Explosion of a giant star.
Tachyon Hypothetical particle that travels at speeds greater than that of light.
Tardyon Particle that travels at speeds less than that of light.
Tensor theory A complex branch of mathematics. The equations of general relativity are written in terms of tensors.
Thermodynamics The study of the dynamics of heat.
Thermonuclear furnace Region in center of a star where nuclear reactions occur.
Tidal forces Stretching forces caused by difference in gravitational pull.
Time dilation A decrease in time interval caused by motion.
Totality Period when solar eclipse is total.
Transformation A change in coordinates. A mathematical relation between two systems.
Twin paradox Paradox created when one twin moves at high speed relative to the other, then returns.
Uncertainty principle Principle that states there is an uncertainty when we attempt to measure various variables in physics simultaneously.
Variable star Star that changes in light intensity.
Virtual pair A particle–antiparticle pair that appears briefly, then disappears.
White dwarf A small dense star slightly larger than Earth.
White hole Associated with the exit end of a black hole.
World line A line giving the position of an observer at various times in the universe.

Bibliography

The following is a list of general and technical references for the reader who wishes to learn more about this subject. References marked with an asterisk are of a more technical nature.

CHAPTER 1: Introduction

Berry, Adrien, *The Iron Sun: Crossing the Universe in Black Holes* (New York: Dutton, 1977).
Earman, John, "On Going Backwards in Time," *Philosphy of Science* (September 1967), 211.
Gardiner, Martin, "Can Time Go Back?" *Scientific American* (January 1967), 98.
Macvey, John, *Time Travel* (Chelsea: Scarborough House, 1990).
Parker, Barry, *Einstein's Dream* (New York: Plenum, 1986).

CHAPTER 2: Einstein and the Elasticity of Time

Bernstein, Jeremy, *Einstein* (New York: Viking, 1973).
Clark, Ronald, *Einstein: The Life and Times* (New York: World, 1971).
Einstein, Albert, *Relativity* (New York: Crown, 1961).
Frank, Philipp, *Einstein: His Life and Times* (New York: Knopf, 1972).
Gardiner, Martin, *The Relativity Explosion* (New York: Vintage, 1976).
Hoffman, Banesh, *Albert Einstein: Creator and Rebel* (New York: Viking, 1972).
Michelmore, Peter, *Einstein: Profile of the Man* (London: Muller, 1963).
Pais, Abraham, *Subtle Is the Lord* (New York: Oxford, 1982).

CHAPTER 3: Space That Bends and Twists

Bell, E. T., *Men of Mathematics* (New York: Simon and Schuster, 1937).
Buhler, W. K., *Gauss: A Biographical Study* (Berlin: Springer-Verlag, 1981).
Hall, Tord, *Carl Friedrich Gauss* (Cambridge: MIT Press, 1970).
Marder, Leslie, *Time and the Space Traveller* (Philadelphia: University of Pennsylvania Press, 1971).
Rucker, Rudy, *The Fourth Dimension* (Boston: Houghton Mifflin, 1984).

CHAPTER 4: Taming the Curvature

Calder, Nigel, *Einstein's Universe* (New York: Greenwich House, 1979).
Clark, Ronald, *Einstein: The Life and Times* (New York: World, 1971).
D'Abro, A., *The Evolution of Scientific Thought* (New York: Dover, 1950).
Gardiner, Martin, *The Relativity Explosion* (New York: Vintage, 1976).
Hoffman, Banesh, *Albert Einstein: Creator and Rebel* (New York: Viking, 1972).
Howard, D., and Stachel, J., *Einstein and the History of General Relativity* (Boston: Birkhauser, 1989).
Pais, Abraham, *Subtle Is the Lord* (New York: Oxford, 1982).

CHAPTER 5: The Nature of Time and Space-time

Davies, Paul, *The Physics of Time Asymmetry* (Berkeley: University of California Press, 1974).
Gold, T., *The Nature of Time* (Ithaca: Cornell University Press, 1967).
Landsberg, P. T., *The Enigma of Time* (Bristol: Adam Hilger, 1982).
Morris, Richard, *Time's Arrow* (New York: Simon and Schuster, 1984).

CHAPTER 6: The Discovery of Space-time Tunnels

Hawking, Stephen, and Israel, Werner, *Three Hundred Years of Gravitation* (Cambridge: Cambridge University Press, 1987).
Howard, D., and Stachel, J., *Einstein and the History of General Relativity* (Boston: Birkhauser, 1989).

CHAPTER 7: Rips in the Fabric of Space: Black Holes

Bartusiak, Marcia, *Thursday's Universe* (New York: Times Books, 1986).

Gribbin, John, *Spacewarps* (New York: Delacortes, 1983).

Kaufmann, William, *Black Holes and Warped Space-time* (San Francisco: Freeman, 1979).

Hawking, Stephen, *A Brief History of Time* (New York: Bantam, 1988).

Parker, Barry, "In and Around Black Holes," *Astronomy* (October 1978), 6.

Parker, Barry, "Black and White Holes: Minis, Maxis and Worms," *Star and Sky* (December 1979), 32.

Shields, Gregory, "Are Black Holes Really There?" *Astronomy* (October 1978), 6.

CHAPTER 8: Spinning Gateways

Boslough, John, *Stephen Hawking's Universe* (New York: Avon, 1985).

Chaisson, Eric, *Relatively Speaking* (New York: Norton, 1988).

Darling, David, "Space, Time and Black Holes," *Astronomy* (October 1980), 66.

Kaufmann, William, *The Cosmic Frontiers of General Relativity* (Boston: Little, Brown, 1977).

Parker, Barry, "Mini Black Holes," *Astronomy* (February 1977), 26.

CHAPTER 9: Searching for Time Tunnels

*Cowley, A., Crampton, D., Hutching, J. B., Remillard, R., and Penfold, J. E., "Discovery of a Massive Unseen Star in LMC X-3,"*Astrophysics Journal* **272** (September, 1983), 6.

Davis, Joel, "The Puzzle of SS-433," *Astronomy* (July 1980), 28.

Hawking, Stephen, and Israel, Werner, *Three Hundred Years of Gravitation* (Cambridge: Cambridge University Press, 1987).

McClintock, Jeffrey, "Do Black Holes Exist?" *Sky and Telescope* (January 1988), 28.

Parker, Barry, "The Other Black Hole Candidates," *Star and Sky* (February 1981), 26.

Shipman, H. L., *Black Holes, Quasars and the Universe* (Boston: Houghton Mifflin, 1980).

CHAPTER 10: Journey into a Black Hole

Cunningham, C. T., "Optical Appearance of Distant Objects Near and Inside a Schwarzschild Black Hole," *Physical Review D* **12**(2) (July 1975), 323.

Kaufmann, William, *The Cosmic Frontiers of General Relativity* (Boston: Little, Brown, 1977).
*Metzenthen, W. E., "Appearance of Distant Objects to an Observer in a Charged Black Hole Space-time," *Physical Review D* **42**(4) (August 1990), 1105.
*Morris, Michael, and Thorne, Kip, "Wormholes in Space-time and Their Use for Interstellar Travel," *American Journal of Physics* **56**(5) (May 1988), 395.

CHAPTER 11: Overcoming the Problems

Freedman, David, "Cosmic Time Travel," *Discover* (June 1989), 58.
*Frolov, V. P., and Novikov, I. D., "Physical Effects in Wormholes and Time Machines," *Physical Review* D **42**(4) (August 1990), 1057.
*Morris, Michael, and Thorne, Kip, "Wormholes in Space-time and Their Use for Interstellar Travel," *American Journal of Physics* **56**(5) (May 1988), 395.
*Morris, Michael, Thorne, Kip, and Yurtsever, Ulvi, "Wormholes, Time Machines and the Weak Energy Condition," *Physical Review Letters* **61**(13) (September 1988), 1446.
Redmount, Ian, "Wormholes, Time Travel and Quantum Gravity," *New Scientist* (April 1990), 57.
*Visser, Matt, "Traversable Wormholes: Some Simple Examples," *Physical Review D* **39**(10) (May 1989), 3182.

CHAPTER 12: Wormholes and Other Universes

*Coleman, Sidney, "Why There Is Nothing Rather Than Something: A Theory of the Cosmological Constant," *Nuclear Physics B* **310** (1988), 643.
*Fischler, W., Morgan, D., and Polchinski, J., "Quantum Nucleation of False Vacuum Bubbles," *Physical Review D* 41(8) (April 1990), 46.
Freedman, David, "Maker of Worlds," *Discover* (July 1990), 46.
*Hawking, Stephen, "Breakdown of Predictability in Gravitational Collapse," *Physical Review D* **14**(10) (November 1976), 2460.
Schwarzschild, B., "Why Is the Cosmological Constant So Very Small?" *Physics Today* (March 1989), 25.

CHAPTER 13: Advanced Civilizations and Time Travel

Abell, George, "The Search for Life Beyond Earth: A Scientific Update," *Extraterrestrial Intelligence: The First Encounter* (Buffalo: Prometheus Books, 1979), 53–71.

McDonough, Thomas, *The Search for Extraterrestrial Intelligence* (New York: Wiley, 1987).
Papagiannis, Michael, "The Search for Extraterrestrial Civilizations: A New Approach," *Mercury* (January–February 1982), 12.
Rood, Robert, and Trefil, James, *Are We Alone?* (New York: Scribners, 1981).
Tipler, Frank, "The Most Advanced Civilization in the Galaxy Is Ours," *Mercury* (January–February 1982), 5.

CHAPTER 14: Faster Than Light?

Freedman, David, "Beyond Einstein," *Discover* (February 1989), 56.
Herbert, Nick, *Faster Than Light: Superluminal Loopholes in Physics* (New York: New American Library, 1988).

Index